NEUROLOGY – LABORATORY AND CLINICAL RESEARCH DEVELOPMENTS

COMAS AND SYNCOPE

CAUSES, PREVENTION AND TREATMENT

NEUROLOGY – LABORATORY AND CLINICAL RESEARCH DEVELOPMENTS

Additional books in this series can be found on Nova's website under the Series tab.

Additional E-books in this series can be found on Nova's website under the E-book tab.

NEUROSCIENCE RESEARCH PROGRESS

Additional books in this series can be found on Nova's website under the Series tab.

Additional E-books in this series can be found on Nova's website under the E-book tab.

NEUROLOGY – LABORATORY AND CLINICAL RESEARCH DEVELOPMENTS

COMAS AND SYNCOPE

CAUSES, PREVENTION AND TREATMENT

EDUARDO SILVA
AND
GABRIEL CRUZ
EDITORS

Nova Science Publishers, Inc.
New York

For permission to use material from this book please contact us:
Telephone 631-231-7269; Fax 631-231-8175
Web Site: http://www.novapublishers.com

Additional color graphics may be available in the e-book version of this book.

Library of Congress Cataloging-in-Publication Data

Comas and syncope : causes, prevention, and treatment / editors, Eduardo Silva and Gabriel Cruz.
p. ; cm.
Includes bibliographical references and index.
ISBN 978-1-62100-603-9 (hardcover)
I. Silva, Eduardo Missel. II. Cruz, Gabriel.
[DNLM: 1. Coma--etiology. 2. Syncope--etiology. 3. Coma--therapy. 4. Syncope--therapy. WB 182]
LC classification not assigned
616.8'49--dc23

2011035667

Published by Nova Science Publishers, Inc. † New York

CONTENTS

PREFACE

In this book, the authors present topical research in the study of comas and syncope. Topics discussed include the effects of acoustic stimulation on comatose patients; cardiogenic syncope; management of non-traumatic comas in sub-Saharan Africa and the role of neuroimaging and genetics in comatose pediatric patients.

Chapter 1 - One of the most important clinical challenges in patients with severe brain damage is to estimate different aspects of their actual state of consciousness and, in particular, their residual conscious perception of the environment. Different methods have been developed to quantify the preservation of responses in the ascending sensory systems and the projection areas, e.g., bedside behavioural assessment and analyses of short-latency event-related potentials.

Most of current research on coma focuses on how to distinguish irreversible from potentially reversible coma but the question of how health professionals should address comatose patients has escaped critical appraisal. Not talking to comatose patients may promote the notion that they are dead or nearly dead, and may influence health professionals to inappropriately withhold or withdraw therapy. Extensive research on the auditory brainstem function of comatose patients shows that the majority have normal brainstem auditory evoked responses. Verbal communication is a way of orientating and providing meaningful sensory input to an unconscious patient. Familiar voices or simply using the patient's name and individualising care may provide better sensory input.

Communication amongst healthcare providers is one of the most important factors associated with quality of care and patient safety, especially in Intensive Care Units (ICUs). However, relatively little is known about the different aspects of communication in health care, specifically in ICUs, and the relation between communication and outcome measures.

Despite its importance, there is evidence that communication in Intensive Care Unit (ICU) settings is not common practice. Inadequate health professional-patient communication results in increased levels of stress and anxiety. Thus, communication within the ICU environment is of paramount importance, and effective communication is an essential aspect of health professionals' care.

One of the most important and challenging tasks for healthcare professionals caring for patients who survive brain injury is establishing a prognosis for long-term functional outcome, as soon as possible after injury, even while the patient is unconscious. Severe brain injury changes many lives instantaneously and prognostication during coma recovery can help with early counselling and adjustment of the patient and family, as well as guide the planning for present and future rehabilitation.

The inconsistencies in the research findings point to the need for further detailed investigation of the effects of voice stimulation on comatose patients.

Based on the literature review the authors developed a study that includes a detailed, objective and systematic analysis of the effects of acoustic stimulation on comatose patients using a standardised stimulus message. The curves of electrocardiography (ECG), blood pressure and pulse plethysmography and the parameters of pulse frequency, systolic, diastolic and mean arterial pressure, oxygen saturation level and peripheral body temperature, were monitored before, during and after the auditory stimulation of the comatose patient with an unknown voice and a familiar voice.

This study contributes to the reflection on the practice of communication with the unconscious patients, in order to sensitise health professionals to the importance of communication in the ICU, to contribute to improving the quality of care and help in establishing the prognosis.

Chapter 2 - In last 15 years, cardiology has experienced an amazing progress in genetics and molecular biology. These advances help to develop new methods of prevention, diagnosis and clinical treatment of cardiac pathologies. The majority of these diseases have an incomplete penetrance; therefore patients may be unaware of their illness. Sudden cardiac death (SCD) is a rare but socially devastating event. In several cases, physical activity can be the trigger for SCD as first symptom. Most common causes of SCD are congenital electrical disorders and structural heart diseases although a significant number of SCD remain unexplained. In the pediatric population, in up to 50% of SCD cases, sudden death (SD) is the first and only clinical manifestation of an inherited cardiac disease that had remained undetected by conventional clinical investigations. Concretely, sudden infant death syndrome (SIDS) accounts for 10% of infant mortality in the first year of life. Most of the SCD in pediatric population are due to channelopathies and cardiomyopathies. This chapter focuses on recent advances in genetic channelopathies and cardiomyopathies in pediatric population and their clinical treatment.

Chapter 3 - Syncope is a transient, brief and sudden loss of consciousness produced by a lessening of the flow of blood to the brain and therefore oxygen deprivation. It can be presented from different ways. Similarly, can be due to a great variety of causes, although the most frequent causes to produce an episode of syncope on any person are: the stroke (embolism and / or brain hemorrhage) and cardiovascular disorders. In many occasions, both are interlinked.

Patients suffer a complete loss of consciousness that may be associated to involuntary movements of the limbs. This may lead to confusion in diagnosis between epilepsy and syncope, while the treatment is quite different for both pathological entities. Unfortunately this confusion is too frequent, so that many cases are treated as epilepsy when in fact is another origin of syncope. Hence, the differential diagnosis should be essential in such events.

All syncope episode gives as result an inexplicable fall of the person who suffers from it.

The most important tool for all physicians for a first diagnostic impression is the patient medical history along with the eyewitness recollection of the incident, if possible. Physicians should be aware of the possibility that a patient or caregiver may inadvertently misinterpret the symptoms when trying to describe to them.

Today, it is well known that approximately 10% of patients diagnosed with epilepsy are syncope sufferers. More than 25% of them have an origin in cardiac abnormalities (this percentage is 45-50% higher when the patient is 60 years or older, regardless of sex).

Any patient with several episodes of syncope or loss of consciousness should be subjected to a full investigation. This should include at least, a blood chemistry analysis, an electroencephalogram (EEF) and an electrocardiogram (ECG), which are fundamental components for evaluation. A normal ECG is a good prognostic feature in most cases.

An ECG without signs suggestive for cardiac pathology should make us think that the syncope is not severe and should be seen as reassuring.

Chapter 4 - Syncope, defined as a sudden, transient loss of consciousness with spontaneous recovery, may be the presenting symptom for conditions ranging from benign illnesses to life-threatening disorders. Common causes of syncope include vasovagal syncope, situational syncope, carotid sinus hypersensitivity, orthostatic hypotension, structural heart diseases, left and right sided obstructive cardiac lesions, and cardiac arrhythmias. History often provides valuable clues about the underlying etiology of syncope. Admission decisions in the emergency department are guided by various risk prediction scores. Evaluation of a patient with syncope involves a large battery of diagnostic tests that include a 12-lead electrocardiogram, Holter monitoring, an echocardiogram, tilt table testing, ischemia evaluation, electrophysiologic studies, and other imaging tests. Despite the availability of these advanced diagnostic tests, a significant proportion of patients with syncope remain undiagnosed. Therapy includes various non-pharmacological measures such as life style modifications, adequate fluid and salt intake, tilt training exercises, and pharmacotherapy with fludrocortisone, midodrine, beta adrenergic blockers, vagolytics, or serotonin reuptake inhibitors. Cardiac pacing may be beneficial in a selected subgroup of patients.

Chapter 5 - This chapter presents the remarkable trajectory of a young boy who experienced an idiopathic acquired brain injury (ABI) of unknown etiology followed by 22 months of coma prior to his gradual road towards recovery. This case is of particular interest because of available premorbid intellectual and achievement assessment data due to a mild pre-existing diagnosed learning disability. As such, it has been possible to compare benchmarks of his ongoing progress to his pre-illness cognitive and educational profile on a yearly basis since 2002 until the present. Previous research indicates that predictions of educational outcome after pediatric brain injury are a challenge. In this case, evidence-based interventions (EBI) are shown to maximize the student's intellectual and psychosocial potential. Although this young man resides in a rural, underserved community, multisystemic interventions have been successfully implemented in order to allow him to continue successfully in his home school district. The provision of multiple psychoeducational services have resulted in steady academic growth, although significant impairment persists in some domains.

The main purpose of this chapter is to examine potential mechanisms underlying the complex relationship between the sequelae of pediatric ABI and intellectual functioning as illustrated by this case study. the authors' secondary goal is to increase awareness of the role of established EBI in the management of patients with ABI at both the individual and the community level.

Chapter 6 - Objective: the aim of the authors' study was to list the main etiologies of non traumatic comas in sub-Saharan Africa and to specify the importance of computed tomography scan in its management.

Patients and method: retrospective analysis of the files of 142 patients hospitalized for non traumatic coma, between January, 2006 and December, 2010, in the neurology, neurosurgery, emergency and intensive care departments of Yopougon Teaching Hospital (Abidjan-Ivory Coast). All the patients had cerebral CT and biological check up.

Results: Patients' mean age was 42 years. The sex ratio was 1.3. The coma was deep (score of Glasgow = 8) in 77.5 % of the cases (n=110). Comas with fever represented 54.9 % of the cases (n=78) and comas without fever represented 45,1 % of the cases (n = 64). HIV - AIDS was associated in 39.4 % (n = 56) of the cases to coma with fever. In 69.7 % (n=99) of the cases, an etiology was found at cerebral CT scan. These causes were represented by cerebral toxoplasmosis (33.8 %; n = 48), ischemic stroke (16.9 % n = 24), brain haematoma (7.8 % n=11), brain tumors (6.4 % n = 9) and subarachnoid hemorrhage (2.1 % n = 3). In 30.3 % (n = 43) of cases the cerebral CT scan was normal and biological etiologic finding were neuro-malaria (10.6 % n = 15), cerebral tuberculosis (4.6 % n = 6), neuro-meningeal cryptococcosis (2.8 % n = 4), diabetic coma (3.5 % n = 5), hyperthyroidism (1.4 % n = 2) and toxic coma (2.1 % n = 3). In 5.7 % of the cases (n = 8) no cause was found.

Conclusion: in Sub-Saharan Africa, the etiologies of comas with fever are dominated by toxoplasmosis cerebral and those without fever are dominated by stroke. The cerebral CT scan helped to find more than the 2/3 of the causes of non traumatic comas and it also helped an immediate management of the pathology.

Chapter 7 - Coma is defined as a profound or deep state of unconsciousness. Comatose patient cannot voluntarily react with the external environment and mostly possess specificaly limited reflex reactivity to external stimuli depending on coma severity. Different levels of coma severity can be determined by neurological examitation based on brainstem reflexes and level of unresponsiveness to external stimuli. Coma severity and further prognosis depend upon the nature and extent of brain dysfunction, determined primarily by causative etiologic factors. Traumatic brain injuries are one of the most common causes of coma. Coma is commonly a result of non-traumatic intracerebral bleeding, ischaemic stroke or inflammatory insult with or without brain swelling. Generally, coma is also result of inadequate oxygen or glucose supply to brain tissue and a great variety of drug and poisons can directly cause a coma too. Initial diagnosis and treatment in comatose patient should start simultaneously. History still remains the basal key to the diagnosis. Complete clinical neurologic assessment, Glasgow Coma Score and clinical physical examination together with electrocardiography, brain computer tomography or magnetic resonance imaging scan, electroencephalography, blood, urine and cerebrospinal fluid laboratory examination should follow in close connection. Immediate treatment is focused to life-threatening issues, as open airway, spontaneous breathing, heart beat and blood pressure, blood glucose level, serum osmolarity and acid-base balance of internal environment. Comatose patients are treated in intensive care units and usually require full life support until their situation resolves in either way. The prognosis and outcome of comatose patient depends on the coma cause, the extent and severity of brain damage, the possibility to correct or normalised affected brain function and comatose state duration. Repeated neurologic assessment is needed to monitoring the coma level. A coma possessing defined clinical features that lasts longer than four weeks is usually reclassified as a persistent vegetative state, sometimes remaining for years. Comas resulting from drug overdosage or poisonings have the good recovery capacity if immediate and proper medical care is received. Generally comas resulting from traumatic brain injuries tend to have a slightly better recovery rate compared to comas related to illness. However comatose patient may never regain consciousness or be permanently seriously disabled.

Vegetative state lasting more than one year has a minimal chance to regain full consciousness. Predicitve factors also include electrophysiology, biochemical and genetic testing regarding evoked potential responses, S 100 protein and apolipoprotein E genotype. Age related specifics need to be considered in comatose pediatric patients.

Chapter 8 - Traumatic brain injury and other neurological disorders are primarily assessed using the Glasgow Coma Scale (GCS) to evaluate consciousness. However, GCS score assignment is far from sufficiently reliable for correct assessment, especially with inexperienced users. In this chapter, the authors reveal what factors of the GCS are misjudged when assessed by inexperienced medical personnel. They analyzed GCS eye, verbal, and motor response (EVM) scoring profiles conducted by postgraduate year 1 junior residents (n = 94) before they began residency in specific departments. GCS assessment was tested using a video simulation that portrayed mock patients with eight different levels of consciousness that are frequently encountered in trauma patients at the emergency department. On average, 26% ± 18% of examinees failed to provide the correct EVM profiles for the eight selected consciousness levels. Primary misjudged GCS factors belonged to two categories: the assessment of “confused conversation (verbal score of 4)”, and the assessment of “withdrawal motor response (motor score of 4)”. Additional instruction regarding the specific misjudged factors described in this chapter may help inexperienced medical personnel improve the reliability of GCS score assignment to casualties or patients with consciousness disturbance.

Chapter 9 - Syncope is a clinical syndrome that is a subset of a broader range of conditions that cause sudden and temporary loss of consciousness, from which recovery is spontaneous.

The causes of syncope can be grouped into four major categories: neurologic, metabolic, vasomotor and cardiac. Of these, only cardiac syncope commonly leads to sudden death.

There are two major groups of cardiac syncope: obstructive cardiac lesions and cardiac arrhythmias. Cardiac arrhythmias represent most common cause and can be subdivided in bradycardias and tachycardias. The aim of this chapter is to provide a clear overview of syncope-producing cardiac arrhythmias.

Chapter 10 - Syncope is a transient and often un-witnessed event with numerous potential etiologies. There is often difficulty in choosing the appropriate diagnostic tests and procedures. Patients with risk factors for cardiac causes of syncope frequently undergo multiple testing modalities with poor yield and in one-third of patients, the cause remains unknown. Diagnosing cardiac causes of syncope is important as it portends a worse prognosis with a high recurrence rate and significant morbidity and mortality. The implantable loop recorder (ILR) is a small device that is implanted subcutaneously and can monitor cardiac rhythm. The ability of ILRs to continuously record cardiac electrical activity over extended periods of time makes them a powerful diagnostic tool. Several studies have now shown that the ILR can be very useful in the diagnostic evaluation of syncope and thus supports the recommendation that an ILR should be implanted early rather than late in the evaluation of infrequent unexplained syncope.

In: Comas and Syncope: Causes, Prevention and Treatment ISBN 978-1-62100-603-9
Editors: Eduardo Silva and Gabriel Cruz. ©2012 Nova Science Publishers, Inc.

Chapter 1

THE EFFECTS OF ACOUSTIC STIMULATION ON COMATOSE PATIENTS

João F. F. L. Simões[1], Luis M. T. Jesus[2], David Voegeli[3], Celso Martins[4], Andreia Hall[5] and David Simpson[6]

[1]School of Health Sciences, University of Aveiro, 3810-193 Aveiro, Portugal
[2]School of Health Sciences and Institute of Electronics and Telematics Engineering of Aveiro, University of Aveiro, 3810-193 Aveiro, Portugal
[3]School of Health Sciences, University of Southampton, Southampton S017 1BJ, UK
[4]Hospital de Santo António, Centro Hospitalar do Porto, 4099-001 Porto, Portugal
[5]Department of Mathematics and Center for Research and Development in Mathematics and Applications, University of Aveiro, 3810-193 Aveiro, Portugal
[6]Institute of Sound and Vibration Research, University of Southampton, Southampton S017 1BJ, UK

1. INTRODUCTION

One of the most important clinical challenges in patients with severe brain damage is to estimate different aspects of their actual state of consciousness and, in particular, their residual conscious perception of the environment (Perrin et al., 2006; Boly et al., 2011). Different methods have been developed to quantify the preservation of responses in the ascending sensory systems and the projection areas, e.g., bedside behavioural assessment and analyses of short-latency event-related potentials (Perrin et al., 2006).

Most of current research on coma focuses on how to distinguish irreversible from potentially reversible coma but the question of how health professionals should address comatose patients has escaped critical appraisal. Not talking to comatose patients may promote the notion that they are dead or nearly dead, and may influence health professionals to inappropriately withhold or withdraw therapy (LaPuma et al., 1990). Extensive research

(LaPuma et al., 1990) on the auditory brainstem function of comatose patients shows that the majority have normal brainstem auditory evoked responses. Verbal communication is a way of orientating and providing meaningful sensory input to an unconscious patient. Familiar voices or simply using the patient's name and individualising care may provide better sensory input (Elliot and Wright, 1999; Holeckova et al., 2006; Perrin et al., 2006).

Communication amongst healthcare providers is one of the most important factors associated with quality of care and patient safety (Hoonakker et al., 2008), especially in Intensive Care Units (ICUs). However, relatively little is known about the different aspects of communication in health care, specifically in ICUs, and the relation between communication and outcome measures.

Despite its importance, there is evidence (Alasad and Almad, 2005) that communication in Intensive Care Unit (ICU) settings is not common practice. Inadequate health professional-patient communication results in increased levels of stress and anxiety. Thus, communication within the ICU environment is of paramount importance, and effective communication is an essential aspect of health professionals' care.

One of the most important and challenging tasks for healthcare professionals caring for patients who survive brain injury is establishing a prognosis for long-term functional outcome, as soon as possible after injury, even while the patient is unconscious. Severe brain injury changes many lives instantaneously and prognostication during coma recovery can help with early counselling and adjustment of the patient and family, as well as guide the planning for present and future rehabilitation.

The inconsistencies in the research findings (Johnson, Omery and Nikas, 1989; Doman et al., 1993; Jones et al., 1994; Walker, Eakes and Siebelink, 1998; Holeckova et al., 2006; Perrin et al., 2006; Puggina, 2006) point to the need for further detailed investigation of the effects of voice stimulation on comatose patients.

Based on the literature review (Simões, Jesus and Voegeli, 2012) we developed a study that includes a detailed, objective and systematic analysis of the effects of acoustic stimulation on comatose patients using a standardised stimulus message. The curves of electrocardiography (ECG), blood pressure and pulse plethysmography and the parameters of pulse frequency, systolic, diastolic and mean arterial pressure, oxygen saturation level and peripheral body temperature, were monitored before, during and after the auditory stimulation of the comatose patient with an unknown voice and a familiar voice.

This study contributes to the reflection on the practice of communication with the unconscious patients, in order to sensitise health professionals to the importance of communication in the ICU, to contribute to improving the quality of care and help in establishing the prognosis.

1.1. Verbal Communication with the Comatose Patient

The ICU can create a great deal of tension and stress for patients and families (Elliot and Wright, 1999). Effective and appropriate communication is an important part of the healing process, not only for the patient, but also for the family.

The ICUs are busy places filled with all types of noise. Patients do not get much sleep, and sleep deprivation along with narcotic drugs can contribute to confusion in the ICUs (so called *ICU psychosis*) due to the critical care environment and illness itself. Many patients become confused just being removed from their homes into a different environment, or in the dark of the night, even when they are not particularly ill. As the patient continues to heal, medication dosage is typically decreased. Normal sleep patterns gradually return and the patient recovers from the confusion of ICU psychosis. The act of communication will aid the patient in the process of returning to normal orientation. Various types and levels of communication will be of benefit at different stages of the recovery process (Alasad and Almad, 2005).

The unconscious patient is under stress because he/she is likely to be misunderstood, and is unsure of what is happening. Furthermore, cognitive reappraisal is virtually impossible in the ICU environment. Verbal communication can reduce stress by facilitating the patient's use of adaptive coping mechanisms (Elliott and Wright, 1999).

Verbal communication is a way of orientating and providing meaningful sensory input to an unconscious patient. Familiar voices or simply using the patient's name and individualising care may provide better sensory input (Elliot and Wright, 1999; Holeckova et al., 2006; Perrin et al., 2006). Hence, verbal communication can help patients preserve self-identity and self-esteem and reduce social isolation (Elliot and Wright, 1999). These factors help to enhance feelings of wellbeing and optimism, a proposed mechanism for decreasing recovery time (Johnson et al., 1989; Doman et al., 1993; Baker and Melby, 1996; Walker et al., 1998; Elliot and Wright, 1999; Holeckova et al., 2006; Perrin et al., 2006).

Communication with a critically ill patient who might be sedated, ventilated, and therefore unable to speak and subject to distorted sensory input, is important (Baker and Melby, 1996; Verity, 1996): it allows access to and exchange of information and emotional support during stress (Elliot and Wright, 1999); it is a crucial means of preventing the feeling of isolation which could lead to the development of ICU psychosis (Verity, 1996).

1.2. Responses of Comatose Patients to Verbal Communication

Research to date has resulted in conflicting findings regarding the effects of stimulation on unconscious patients (Johnson, Omery and Nikas, 1989; Walker, Eakes and Siebelink, 1998). There is, however, sufficient evidence to support the idea that increased mentation and emotional arousal may affect the unconscious patient. Evidence reported in the literature (Johnson, Omery and Nikas, 1989; Jones et al., 1994; Walker, Eakes and Siebelink, 1998) also clearly shows a correlation between the auditory stimulation and increases in arterial blood pressure, pulse rate, respiratory rate, intracranial pressure, body movement and facial movement.

We also found evidence in the literature (Johnson, Omery and Nikas, 1989; Doman et al., 1993; Baker and Melby, 1996; Walker, Eakes and Siebelink, 1998; Elliot and Wright, 1999; Holeckova et al., 2006; Perrin et al., 2006) that unconscious patients have neurological and physiological alterations with auditory stimulation, and that most intensive care health professionals claim that verbal communication with unconscious patients is very important (Ashworth, 1987; Turnock, 1991; Baker and Melby, 1996; Elliot and Wright, 1999; Hoonakker et al., 2008), but there is still some ambiguity as to the unconscious patient's level of awareness.

Major factors influencing communication are the patient's level of consciousness, the amount of physical care being given and the presence of relatives (Baker and Melby, 1996).

In the present study we found that the purpose of verbal communication by critical care health professionals focuses on professional aspects, and the purpose of patients' family communication focuses in personal aspects.

Evidence (Walker, Eakes and Siebelink, 1998) suggests that the majority of comatose patients have normal brainstem auditory evoked responses and indicates that they may hear. Indeed, LaPuma et al. (1990) have gone so far as to suggest that talking to comatose patients may have considerable therapeutic value.

1.3. Purpose of Verbal Communication by Health Professionals

Communication is the foundation of nursing and the art of caring holistically for patients', and much of health care is communication-centred (Verity, 1996). This becomes even more relevant for the unconscious patient who is particularly dependent on hearing for sensory stimulation. Health professionals give directions, offer reassurance, provide consolation, commiserate, interpret and receive information. Therefore, the most effectively and efficiently they communicate, the more accomplished they will become in fulfilling their role as health professionals.

There are a limited number of studies (Ashworth, 1987; Baker and Melby, 1996; Elliot and Wright, 1999) that investigated what health professionals actually say to their patients. However, studies to date suggest that intensive care health professionals are not engaging enough in verbal communication (Elliot and Wright, 1999), tend to concentrate more on the technical aspect of their work, and often fail to meet patient's psychological and social needs by insufficient and/or ineffective communication (Baker and Melby, 1996). Communication with unresponsive patients is limited, relating primarily to nursing interventions (Ashworth, 1987). There appear to be four main problem areas (Baker and Melby, 1996): basic difficulty in communicating with a patient who cannot respond; pressures of the working environment; limited knowledge about unconscious patients' needs; limited detailed knowledge of why or how to communicate with unconscious patients.

Increasing numbers of patients report their experiences, when they recover from coma. A common suggestion by these patients is the need for effective verbal communication (Elliot and Wright, 1999). Unconscious patients have a considerable need for information and support, and verbal communication is a way of orientating and providing meaningful sensory input to these patients.

Research undertaken by Ashworth (1987) showed that the communication aspect of care by intensive care nurses was less skilled than other aspects of their care. This theory was supported by Turnock (1991) who found that nurses did not provide adequate verbal and non-verbal communication. Also Elliott and Wright (1999) concluded that intensive care nurses may not be reflecting and understanding the importance of communication in their practice.

Baker and Melby (1996) concluded that at times verbal communication with unconscious patients was so minimal, that any potential benefit to the patients would have been negligible. Survival is a major consideration in intensive care (Ashworth, 1987), and communicating with the patient may become a low priority, whilst the health professional attends to the demands of highly technical equipment needed to support life and aid recovery (Verity, 1996).

Without adequate communication skills health professionals, can neither assess, plan, implement, nor evaluate care effectively. To achieve effective communication, especially in intensive care, where it is often vital to convey the essence of a message to the recipient in a language that they can understand, requires knowledge of how to convey a message, at the right level, without the use of jargon (Verity, 1996). This is a skill that needs to be taught theoretically and practically, so that barriers to communication may be minimised.

It is the duty of intensive care health professionals to ensure that they provide care and attention which cause the least possible discomfort and distress to patients (Baker and Melby, 1996). Ashworth (1987) suggested through friendly interaction, giving relevant information and encouraging the patient to communicate (if possible), health professionals could help to prevent such adverse effects.

It is the health professionals' responsibility to promote all avenues of communication in order to lessen the effects of the communication deficit, thereby promoting patient orientation which could reduce the risk of ICU psychosis. The necessity for reassurance, encouragement and reorientation is very important when the patient is deprived of some or all methods of communication. Therefore, formal support systems and continued education of health professionals about the benefits of verbal communication is deemed necessary.

1.4. Purpose of Verbal Communication by Patients' Family

Research about verbal communication by the patients' family has also resulted in conflicting findings. While families' visits to the ICU have been thought by some to precipitate detrimental changes in physiological variables such as heart rate and blood pressure (Walker, Eakes and Siebelink, 1998), others have observed no untoward effects associated with such stimulation. Still, opportunities for family interaction with comatose patients are usually very limited.

Familiar voices or simply using the patient's name and individualising care may provide better sensory input (Elliott and Wright, 1999). Jones et al. (1994) examined physiological measures (pulse rate, respiratory rate, body movement, and facial movement) using four different auditory stimuli (rock music, classical music, nature sounds, and family/friend voices). Their (Jones et al. 1994) findings suggested that taped voices of family and friends consistently resulted in greater increases in arousal than did other types of taped stimuli.

The study of Walker, Eakes and Siebelink (1998) revealed no negative effects associated with exposure to taped familial voices. However, no significant changes were evidenced in intracranial pressure (ICP), blood pressure, pulse, respiratory rate, mean arterial pressure, oxygen saturation level, or level of restlessness for any of the study participants. No experimental intervention had to be stopped because of an adverse reaction. The study (Walker, Eakes and Siebelink 1998) findings, therefore, add support to allowing family interactions with comatose head-injured patients during their hospitalisation in ICU settings.

2.Method

2.1. Aims

The primary aims of this study were: to understand the effects of auditory stimulation on comatose patients' physiological parameters and monitored curves; to explore the differences between the use of an auditory stimulus based on an unknown voice versus a familiar voice; to study the impact of this data on the prognosis of Coma.

2.2. Ethical Considerations

The Hospital's Ethics Committee gave full ethical approval and the study was registered with the Hospital's Research Office, thus fulfilling local research governance requirements. Meetings were held with ward staff prior to the study commencing. Ward staff contacted patients' relatives, so they would consent to be part of the study. All participants were given written information about the study and time to decide whether they wished that the patient take part, and they all gave written consent. Relatives were assured that there was no obligation to take part and that the care of the patients would not be affected should they decline. All data were confidential and kept securely in locked filing cabinets and password protected computers.

2.3.Participants

The study was carried out in a Medical/Surgical adult ICU (Serviço de Cuidados Intensivos 1 do Hospital de Santo António, Porto, Portugal).

Considering the temporal issues, the clinical characteristics of the population in study, the intention to reduce the variables that could affect the results and finally the data of the patients admitted to the ICU during the year of 2006, the sample consisted of10 comatose patients, which were selected using the following inclusion and exclusion criteria:

- Inclusion: Loss of consciousness; Absence of macroscopic response to auditory stimulation; Coma diagnosis; Total score of the Glasgow Coma Scale (Jennett, 2002) between 6 and 9; Aetiology medic, surgical or politrauma without brain injury; Hospitalised for more than 4 days; Identical care assistance; Identical environmental conditions.
- Exclusion: Hearing loss; A score higher than0in the "Hearing" sub-scale of the IARCUA (Simões, Jesus, Voegeli, Couto, Fernandes and Morgado 2011); History of excessive consumption of alcohol and/or narcotic substances; Epilepsy or degenerative diseases of the central nervous system; Diagnosed injury of the auditory pathways and the cerebral cortex; Significant changes of the auditory evoked responses; Induced Coma by infusion of anaesthetic medication.

2.4. Selection of the Social Group and Significant Person

As mentioned previously, the comatose patient was stimulated acoustically with a recorded message by an unknown person and the most meaningful person to him. Thus, we held a meeting with the significant people of the comatose patients to explain the purpose of the study and how they could collaborate. In this explanation we indicated that only one (significant) person could record the stimulus message. The following inclusion and exclusion criteria were defined:

- Inclusion: Belonging to the same social group as the comatose patient, i.e., having shared objectives, values, activities, characteristics and meanings, interrelating with each other during an extended period of time in a process of continuous communication and interaction, interdependence of individuals which is governed by a set rules or common standards, having group consciousness (Bolander, 1994; Prados, 2000; Stanhope and Lancaster, 2011); Accepting to participate in the study; Having a psychological profile to cooperate in interviews and recordings; Communicate verbally; Portuguese language speaker; Being able to read.
- Exclusion: No ability to communicate verbally.

Once selection of people had occurred, individual meetings were held with each significant person to apply sociometric tests. At this meeting we asked two questions. Firstly we asked the person to identify a list of people that would be meaningful to the comatose patient and who would constitute their social group. Secondly we asked them to elect the people who they considered most significant, so that a sociomatrix with only a reasonable number of participants could be built.

After the formation of the social group, we used sociometric techniques (Bastin, 1980; Kerlinger and Lee, 2000) to determine the member most significant for the comatose patient. The method consists of the description and evaluation of social situations and social processes by measuring the proportion of mutual acceptance and rejection observed among individuals in a group. This is determined through a survey of group members about who the comatose person would most or least like to have as company during particular activities (Kerlinger and Lee, 2000). In our case we just set out to assess the acceptability of the elements of the social group.

The aim was that individuals would place the previously identified people in descending order of importance to determine who was most important to the comatose patient. Each person identified as being a member of the social group of the comatose patient was informed of the objectives and relevance of the study, and what taking part would involve. They were asked for permission to access and use information provided by them and guaranteed confidentiality of all data. They then gave their informed consent to participate in the study. Then the question was made and their response recorded.

After collecting the data from each member of the social group, they were organised into a sociometric matrix or sociomatrix (Kerlinger and Lee, 2000) using the program *Sociogram* (Adit, 2006). We used the first 3 people prioritised by each participant to build our matrix. The numerical identifiers of each member of the social group were entered into the first column with row 1 being assigned to the comatose patient as seen in Figure 1. The numerical identifiers of the selected first 3 choices were then placed in the subsequent columns.

The *Sociogram* program then calculates a score corresponding to each choice, allocating maximum marks to to the first choice and minimum marks for the third choice. These scores were then used to build sociograms which synthesize the data and present them visually (see Figure 2). Each sociogram then represents the sociometric status of each individual in the group by plotting the inter-relations between the people (Bastin, 1980).

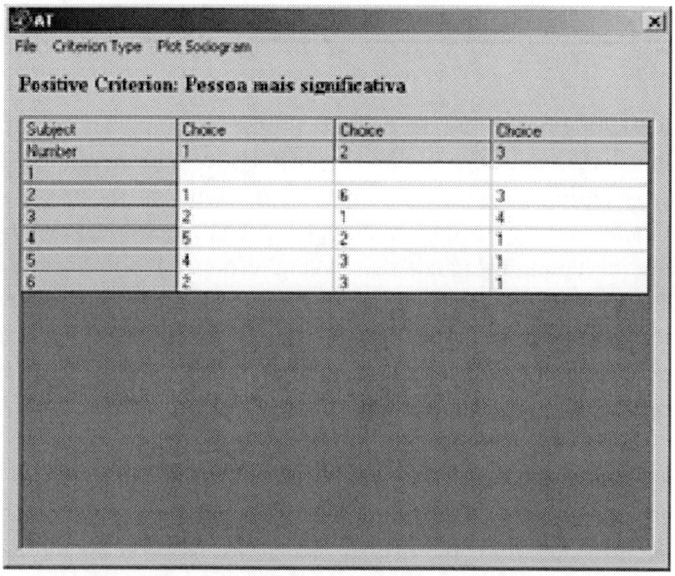

Subject Number	Choice 1	Choice 2	Choice 3
1			
2	1	6	3
3	2	1	4
4	5	2	1
5	4	3	1
6	2	3	1

Figure 1.Sociomatrix of the social group of the comatose patient AT.

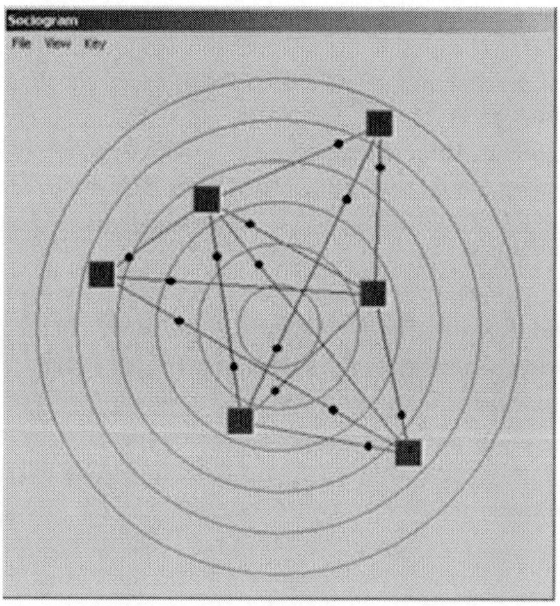

Figure 2. Group sociogram preferences of the comatose patient AT.

The sociograms (see Figure 2) consist of concentric circles where the inner circle represents those closest to the comatose patient and the outer circle those with the lowest scores and therefore more distant. The relationships between the members of the group are represented by the black lines linking the person who chooses and the person chosen. A black dot is placed closest to the person who chooses. If both members mutually selected each other two black dots are used.

2.5. Stimulus Message

2.5.1. Construction of the Stimulus Message

In this study an initial stimulus message was used as a template to construct the stimulus message of the unknown and significant person. These messages were recorded and used later to stimulate the comatose patients.

The stimulus message was based on the results of a content analysis of the verbal communication of critical care nurses and patients' families, as reported in the literature (Simões, Jesus and Voegeli, 2012), and the cooperation of a group of experts (speech and language therapists, and psychologists).

The content analysis of papers about verbal communication with unconscious patients was performed by examining the relative frequencies with which thematic areas, categories and sub-categories occurred. The data were analysed manually using the framework approach (Ritchie and Spencer, 1994; Fischer, 2006; Anderson, 2007), combining pre-established themes with themes arising from the data to develop a coding framework. As with other qualitative research methods, analysis involved abstracting salient features from the immense detail of raw data. Re-reading, clustering and condensing the data lead to the emergence of a number of sub-categories and categories.

The thematic areas that were selected, attempt to justify the importance of the communication with the unconscious patients and also justify the content of the verbal communication of the intensive care health professionals and the patients' relatives.

The expert group proposed the construction of a message that would have an increasing degree of stimulation throughout its deployment. It was judged important that the stimulus, should include both pleasant contents (e.g., "Your family told me they really like you and that they wish you recover quickly") and orders (e.g., "Mr./Mrs. A, I'm here to help you, come on, open your eyes"). This order was thought to be more likely to provoke an exacerbated reaction in the stimulated patient.

The stimulus message consisted of three parts: Presentation and orientation; Information; Functional assessment and stimulation (see Figure 3).

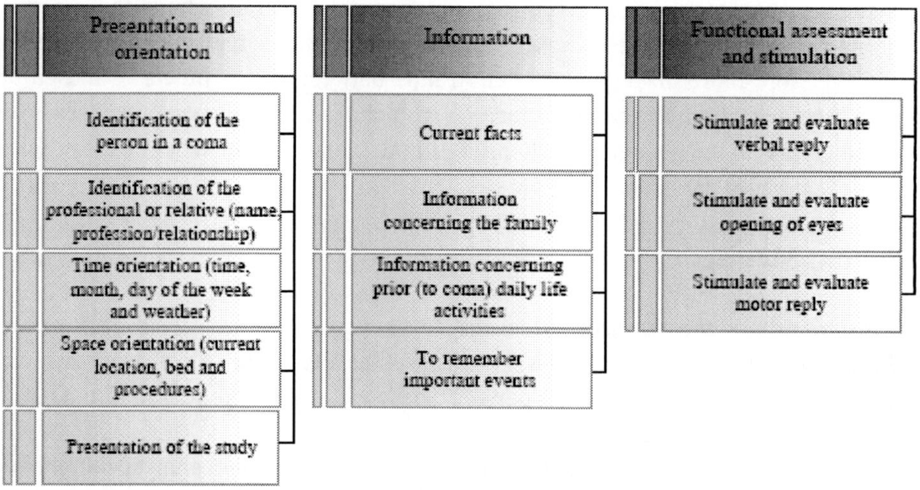

Figure 3.Composition of the stimulus message. Adapted from Simões, Jesus and Voegeli (2012).

The first part included contents with the purpose of introducing the speaker and orientating the conversation in space and time. This part was composed of the acknowledging the person in coma, identifying the health professional or relative (name, profession or relationship), stating the date, day of the week and the weather, and explaining their current location. Finally the study was presented to the patient.

The second part consisted of providing the patient with information about: Current affairs; Family; Daily life activities prior to coma; Important events.

The last part of the stimulus message had the purpose of assessing the functional abilities and response to stimulation of the patients. This part was composed of instructions (e.g., open your eyes) and an evaluation of their responses, including verbal replies, opening of eyes and motor responses.

2.5.2.Recording and Editing the Stimulus Message

After identifying the significant people we explained, to each, the process of recording the stimulus message and how they could participate. Informed consent was required for them to be recorded whilst reading the stimulus message. Then a short interview was conducted to collect data about the experiences of the comatose patient and the names by which each significant person called him. These data enabled the adaptation of message to each individual patient. The final result was validated by the significant person to verify whether the information was entered properly.

Next, the recording of the message was performed in a room of the ICU that provided as much privacy and silence as possible.

Subsequently we proceeded to the choosing an unknown person. This selection was made by a convenience method meeting the following criteria: Accept to participate in the study; Portuguese speaking; Same gender as the significant person; Being unknown to the comatose patient; Not having had any contact with the comatose patient since he had been admitted to the ICU. The stimulus message of the unknown person was based on that of the significant person, and recorded in a room at the University of Aveiro which had identical conditions to the room where the recording of the significant person was conducted.

The recordings of the stimulus messages were performed with a directional microphone Shure PG 48, located 20 cm away from the mouth and in front of the person recorded. The acoustic signals were collected using an external sound card, Edirol-UA-25 to 16 bits and with a sampling frequency of 16 kHz, saved to a .wav file and edited with *Audacity 1.2.6* (Bland et al., 2006).

During the editing process, we avoided changing the content of the message and attempted to make the stimulus as natural as possible, with no repetition of parts of the written text. The editing process was based only on the insertion of pauses where necessary, clipping the messages when too long, removing mistakes and removing emotional reactions which were considered more extreme (e.g., crying).

2.6. Data Collection

The process of data collection began by explaining the study aims to the comatose patient and providing information about what would happen. The sample characteristic questionnaire, which collected demographic and clinical data, was then completed. The monitoring devices were placed correctly and we ensured that they were able to accurately record, and display on the multiparameter monitor all the required parameters for the study.

Before presenting the auditory stimuli to the comatose patient we ensured that they were haemodynamically stable and verified that they had not been subjected to any medical or nursing intervention in the previous 10 minutes. We also gained the agreement of the multidisciplinary team not to perform any clinical interventions during the stimulation, except in emergencies. In this case, we would repeat the stimulation later, once the patient stabilised.

A laptop computer was placed beside the comatose patient. This was used to: Produce an auditory alarm 30 seconds preceding the auditory stimulation; Generate the two stimuli; Collect the various parameters selected previously. The laptop was connected to the multiparameter monitor via an interface cable Datex-Ohmeda ADU-AS/3 Monitor Cable. Headphones (Roland RM-50) were placed on the comatose patient and connected to the internal sound card (SoundMAX Integrated Digital Audio) of the laptop computer. The stimulus message was played using the *Audacity 1.2.6* program (Bland et al., 2006). The researcher simultaneously listened to the playback of the stimulus message to the comatose patient, to make sure that the optimal playback was achieved. The playback volume used for all comatose patients was always the same.

We chose to divide the collection of data into two different time slots (stimulation with a familiar voice and stimulation with the unknown voice) each lasting a total of 45 minutes (see Figure 4).

The following data were monitored during auditory stimulation: Curves of ECG; Invasive arterial pressure; Pulse plethysmography; Pulse frequency; Blood pressures (systolic, diastolic and mean); Oxygen saturation level; Peripheral body temperature.

The time interval between stimulation with an unknown voice and a familiar voice was always at least 30 minutes. We considered this to be the minimum possible whilst allowing for the necessary patient care to be provided.

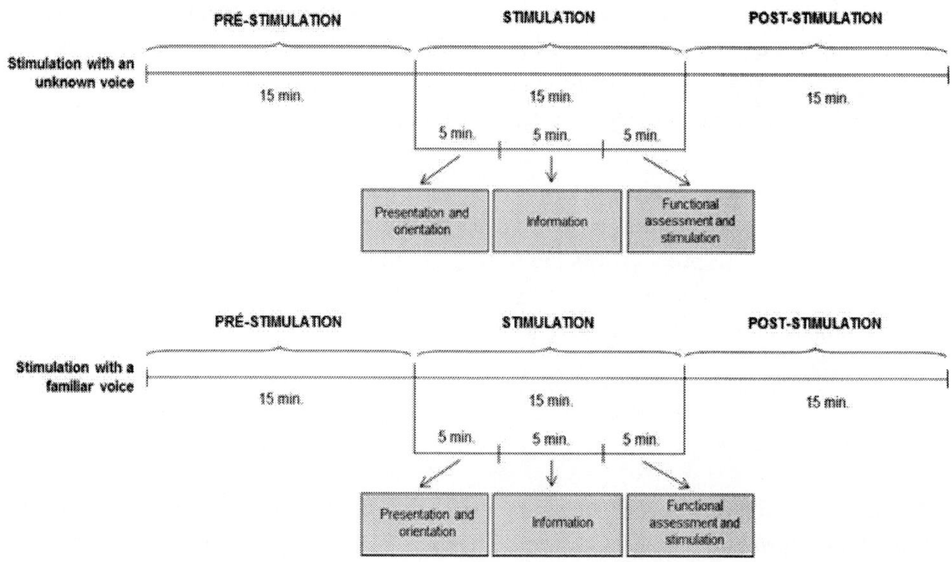

Figure 4. Auditory stimulation.

The stimulation and data collection was always conducted at the end of the afternoon shift. This was the quietest time of day and did not interfere with the patient's biorhythm (e.g., sleep) or relatives visiting times. Selecting a consistent time for stimulation also reduced variability in data which might have occurred due to fluctuations in the patients biorhythms.

The sampling frequency was 0.2 Hz for the values of the parameters and 300 Hz for the curves monitored. In the study we used a multiparameter monitor of the ICU (*Compact Critical Care Monitor* from *Datex-Ohmeda).* This was connected to the computer via a Datex-Ohmeda ADU-AS/3 Monitor Cable (Datex-Ohmeda, 2003).

The COM1 port for data export was selected and the program started. Data were always recorded in DRC format (.drc) and converted to ASCII (.asc) format.

2.7. Data Analysis

2.7.1. Analysis of the Monitored Parameters

The ASCII files of the monitored parameters were imported into the *Excel 2003* program to create individual databases for each comatose patient. This program enabled us to construct various plots for visual interpretation of the parameters monitored during the three periods of stimulation and the three phases of the stimulation message with an unknown voice and a familiar voice.

After this initial analysis, the data were imported into the *SPSS 17.0* program, constructing two databases: one for the periods of stimulation and other for the phases of stimulation. We calculated the minimum and maximum values, the mean and the standard deviation for each monitored parameter during the periods of stimulation with an unknown voice and a familiar voice and in each phase of the stimulus message (Johnson et al., 1989; Walker et al., 1998). This analysis aimed to investigate whether there were significant differences in these values in the different periods and phases mentioned.

We could not apply the commonly used statistical tests (parametric and nonparametric) to the complete data set because the samples are not random. That is, the observations made for each comatose patient are dependent throughout the periods and phases, and to measure the variability we need independent observations. So, we decided to build a new database in *SPSS* with all the minimum and maximum values, means and standard deviation of all the comatose patients, grouped by monitored parameter.

For the minimum, maximum and standard deviation we chose to apply the Friedman test (nonparametric), because the small sample size of our data did not allow us to perform normality tests (Kerlinger and Lee, 2000). For the values of the means we applied a simple analysis of variance (parametric test) because based on the Central Limit Theorem, we can say that the distribution approaches normality (Murteira et al., 2007).

These tests enabled us to determine whether there were any statistically significant differences between the three periods of stimulation and the three phases of the stimulus message. In this exploratory study, with such a small sample size, we used the statistical significance level $\alpha = 10\%$. We chose this value in order to increase the power of the tests which are strongly affected by the small sample size (Murteira et al., 2007).

Where differences were statistically significant, we performed multiple comparison tests: Wilcoxon test for the minimum, maximum and standard deviation; t tests for the mean values. In all cases we applied the Bonferroni correction, i.e., the significance level α was divided by 3.

After this process we also applied the Wilcoxon test to the minimum, maximum and standard deviation, and the t test to the means, now to identify significant differences between the stimulation with an unknown voice and a familiar voice during the periods of stimulation and post-stimulation and also for the three phases of the stimulus message.

We have chosen only to compare between the periods of stimulation and post-stimulation (see Figure 4), as the comatose patient had not been stimulated at baseline. The literature shows some evidence that comatose patients react more significantly to familiar voices while they are stimulated (Hoyt 1996; Walker et al., 1998; Alasad and Ahmad 2005). Furthermore, unilateral tests are recommended when it is expected to have results in one specific direction (Murteira et al., 2007). Therefore, unilateral tests were applied, as we initially expected that the comatose patients would react more relevantly to the familiar voice.

According to our knowledge, there are no studies which have applied stimulus messages composed by different phases. Therefore, for the phases of the stimulus message (see Figure 4), we used a bilateral analysis, because we did not know if the reaction would be further exacerbated by stimulation with a familiar or an unfamiliar voice.

Next, we conducted a separate analysis of each parameter monitored for each comatose patient to check if there were clinically relevant differences in each participant. This analysis was based on the methodology used by Johnson, Omery and Nikas (1989).

Variations in the parameter values were analysed separately in terms of their clinical relevance using the values shown in Table 1. These values were based on literature review (Bolander, 1994; Urden, Stacy and Lough, 2010), the professional experience of the researchers and validated by a group of experts in the field of Intensive Care Nursing.

Table 1. Reference values for the analysis of clinical relevance of the monitored parameters. A variation from previous values was considered clinically relevant when the values exceeded the range presented (for minimal and maximum values analysis). Mean and standard deviation analysis of clinical relevance was based on half the range shown

Monitored parameters	Reference values
Pulse frequency	±10.0 p/m
Systolic blood pressure Diastolic blood pressure Mean blood pressure	±10.0 mmHg
Peripheral body temperature	±1.0 °C
Oxygen saturation level	±5.0%

2.7.2. Analysis of the Monitored Curves

The data of each monitored curve were exported to an ASCII file (.asc) and analysed individually. The duration of RR intervals were calculated and the differences in the three periods and phases of stimulation were analysed. Figure 5 shows the graphical representation of this interval, which is related to the pulse frequency of the comatose patient. This allowed us to verify if the pulse frequency increased or decreased with the auditory stimulation and to compare the analysis results of the ECG curve with the analysis of pulse frequency reported previously.

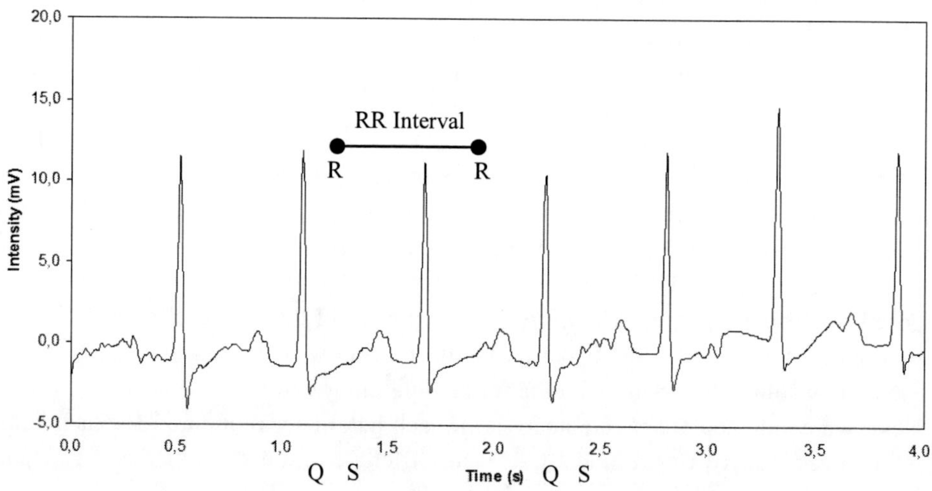

Figure 5.Graphical representation of the RR interval of the ECG curve of the participant AT during the stimulation with an unknown voice.

For the analysis of RR intervals we used a tool developed specifically for this purpose (*check_signals.m* program) using the program *Matlab R2007b* to calculate the intervals between R peaks of QRS complex.

With this tool, the heartbeats are detected by a simple thresholding method (threshold manually selected), applied to the band-pass filtered signals (3-15 Hz). It should be noted that depending on ECG morphology, either positive or negative-going peaks are used.

The isolated peaks (maximum two QRS complexes) in the RR interval time-series were visually inspected, and removed. They were first replaced by 'not a number' (NaN). The raw RR-interval time-series was then linearly interpolated at 4 Hz, and any missing heartbeats were replaced also by linear interpolation between adjacent RR intervals.

The output resulting from this review process was stored in the format .mat, accompanied by a file with the times in which several R peaks were detected. Thus, we had two files for each comatose patient (one for the unknown voice and another for the familiar voice), each with a different number of samples. These data were then imported to the *Excel 2003* program. We also constructed plots for a visual interpretation of the presence or absence of changes in RR intervals during the periods of stimulation with an unknown voice and a familiar voice.

After this initial analysis, the data were imported into *SPSS 17.0*. We then calculated the minimum and maximum values, means and standard deviations (Johnson et al., 1989) of the values of RR intervals during the periods of pre-stimulation, stimulation and post-stimulation with an unknown and a familiar voice. The same procedure was applied to the phases of the stimulus message.

The statistical analysis process followed the guidelines outlined above for the analysis of the parameters with the level of statistical significance set at a maximum of 10%.

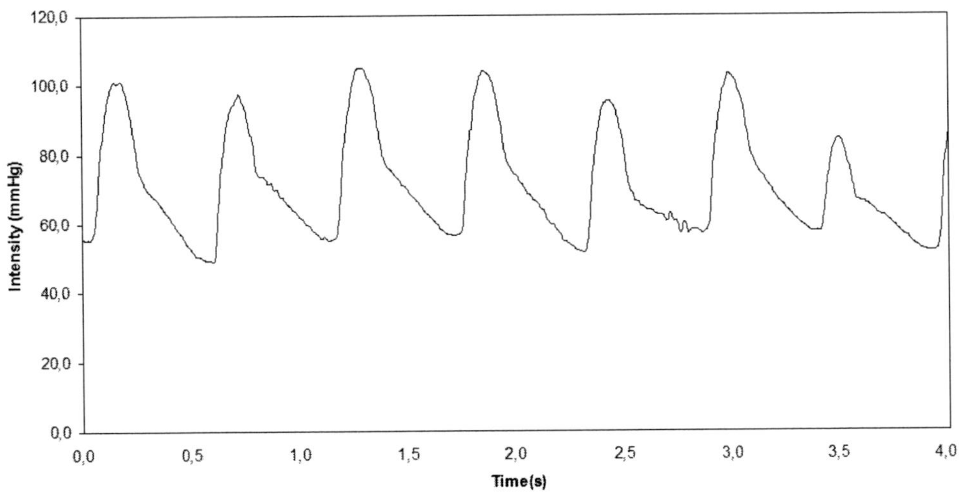

Figure 6.Graphical representation of the Arterial Pressure curve of the participant AT during the stimulation with an unknown voice.

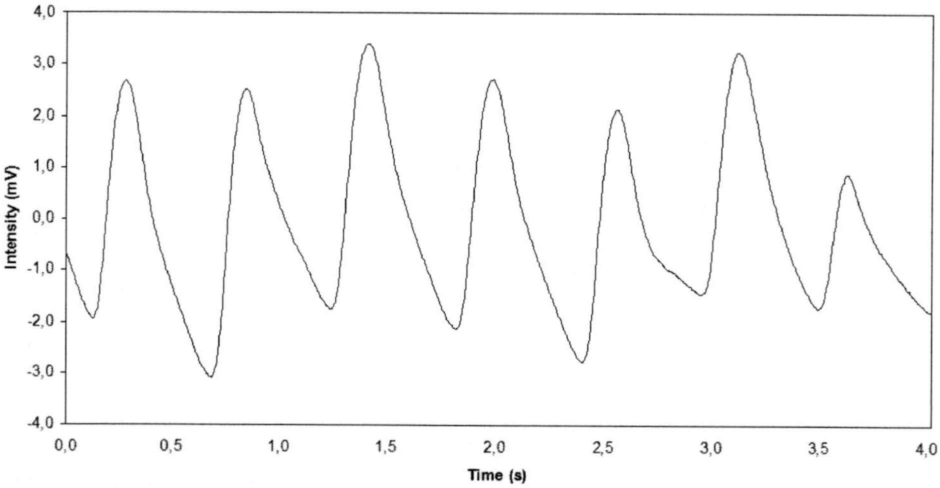

Figure 7.Graphical representation of the Pulse Plethysmography curve of the participant AT during the stimulation with an unknown voice.

Next, we conducted a separate analysis per comatose patient to check whether there were any clinically relevant differences in each participant.

In relation to the analysis of the curves of Blood Pressures and Pulse Plethysmography (represented in Figures 6 and 7, respectively) we conducted research into the potential of analysing these two curves (Allen, 2007; Rangayyan, 2002). This was performed in *Matlab R2007b*. As we had the aim of applying the same statistical tests, before mentioned on the analysis of the ECG wave, we chose to calculate the means and standard deviations in the different periods and phases of stimulation for each comatose patient and type of stimulation (unknown and familiar voice). In this case we did not calculate the minimum and maximum values, because a visual analysis of the graphical representations obtained in *Matlab R2007b*, identified artefacts in the signal, resulting in maximums and minimums that could lead to misinterpretation of results.

Table 2. Reference values for the analysis of clinical relevance of the monitored curves. A variation from previous values was considered clinically relevant when the minimum and maximum values exceeded the range presented (for minimal and maximum values analysis). Mean and standard deviation analysis of clinical relevance was based on half the range shown

Monitored curves	Reference values
RR intervals	±0.05 s
Arterial pressure	±5.0 mmHg
Pulse plethysmography	±0.05 mV

After obtaining the means and standard deviations for the periods and phases of stimulation, two databases were built in *SPSS 17.0* with the aim of applying the Friedman test in the case of standard deviation values and a simple analysis of variance (parametric test) in the case of the mean values. The remaining data analysis followed the same guidelines described above in relation to the ECG curve.

The clinical relevance of the variations in all values derived from the analysis of monitored curves was based in the values shown in Table 2. These values resulted from an extensive literature review (Allen, 2007, Castells et al., 2007, Clifford, Azuaje and McSharry, 2006; McSharry and Clifford, 2006; Oliveira, Maojo, Martin-Sanchez and Pereira, 2005; Rangayyan, 2002; Reisner, Clifford and Mark, 2006; Saritha et al. 2008; Simoliuniene et al. 2008; Urden et al., 2010), the professional experience of the researchers and were validated by a group of experts in Intensive Care Nursing.

3. Results

3.1. Demographic and Clinical Characteristics of the Sample

Four participants were eliminated after the start of the study because their clinical conditions worsened during the period of data collection, meaning they no longer met the study inclusion criteria. One participant was excluded from the study due to technical failures of the software used for the collection of monitored parameters and curves. Also one of the participants was excluded from the study because he was transferred out of the ICU during the data collection period. Only one potential participant was excluded because of changes in auditory evoked responses. Table 3 shows the characteristics of patients included in our study.

In our study the sample consisted of 10 patients in coma (8 male and 2 female). Their ages ranged between 20 and 79 years with a mean age of 56 years.

Table 3. Demographic and clinical characteristics of patients included in the study (*n*=10)

Characteristics	*n*	Mean ± SD	Range
Age (years)	10	56.0 ± 18.4	20-79
Gender			
Males	8		
Females	2		
Days in coma	10	8.8 ± 3.9	3-15
IARCUA total score	10	3.2 ± 0.6	2-4
GCS total score		6.8 ± 0.8	6-8
6	4		
7	4		
8	2		
9	0		
Aetiology			
Medical	6		
Emergency Surgery	3		
Transplant	1		

The number of days in coma ranged from 3 to 15 days, with an average of 8.8 days. This value is relatively low, given the urgency of collecting data due to the clinical instability of these patients. Regarding the aetiology six patients were hospitalised for medical reasons (e.g., Hepatic Coma and Sepsis), three for surgical emergencies and one patient following complications after a transplant.

The total IARCUA scores ranged from 2 to 4 and the total score of Glasgow Coma Scale from 6 to 8.

3.2. Statistical and Clinical Analysis of the Monitored Parameters

3.2.1 Statistical and Clinical Analysis in the Periods of Stimulation

The results of the Friedman and ANOVA tests for the parameters monitored during the three periods of stimulation with an unknown voice are presented in Table 4.

Assuming $\alpha = 0.10$ only the differences in minimum and maximum values of peripheral body temperature and the standard deviations of pulse frequency were statistically significant.

In these three cases we performed multicomparison tests as described previously. However, in all cases, we concluded that there were no statistically significant differences in the comparisons made, despite the Friedman test being significant when we clustered the three periods. This contradiction in results may be due to the high value of the significance level (increasing the probability of a Type I error in the first test) or to the small sample size (increasing the probability of a Type II error in the multiple comparisons test) or eventually to the existence of ties (repeated values).

For minimum values of the whole sample and from a clinical point of view, we found that the pulse frequency values remained very stable in the transition of the periods of data collection. This relative uniformity in the values of the monitored parameters was also found for the other parameters, with no relevant differences from a clinical point of view.

Table 4.Statistical analysis of the parameters monitored during the periods of pre-stimulation, stimulation and post-stimulation with an unknown voice (p-values of the various tests)

	Friedman test			ANOVA
Parameters	p-values of the minimum value	p-values of the maximum value	p-values of the standard deviation	p-values of the mean
Pulse frequency	0.227	0.382	**0.081**	0.187
Systolic blood pressure	0.897	0.601	1.000	0.864
Diastolic blood pressure	0.316	0.314	0.648	0.896
Mean blood pressure	0.897	0.222	0.437	0.914
Peripheral body temperature	**0.038**	**0.017**	1.000	0.172
Oxygen saturation level	0.142	0.899	0.396	0.974

Bold – Statistically significant (p-value < 0.10).

Analysis of the maximum values of the pulse frequency revealed a trend for these values to vary between periods, although this was not statistically significant ($p = 0.382$). So, from a clinical point of view, the maximum values of the whole sample increase between the pre-stimulation period and stimulation periods and decrease again when the stimulation ends.

We identified a clinically relevant change in the minimum and maximum values of the maximum systolic blood pressure in the transition between the periods of pre-stimulation and stimulation although this was not statistically significant. This increase could be associated with anxiety and stress reported by Alasad and Ahmad (2005) and Doman et al. (1993), or the excitatory effect of hearing someone speaking. When the stimulation ends, we noted a marked decrease in the minimum values. The mean values seem to decrease in the transition between the three periods of stimulation. The values tended to normalise when the auditory stimulation finished, returning to pre-stimulation baseline.

For the clinical analysis of the standard deviations resulting from the initial analysis of all the comatose patients of our sample, we can conclude that there were relevant differences in values for the pulse frequency between the period of pre-stimulation and the period of stimulation (minimum, maximum and mean values). After stimulation values decreased, returning to the values obtained before stimulation. This is in accordance with the Friedman test performed previously (we stated that there were significant differences between the three periods of stimulation). This increase in the values of standard deviation indicates the existence of a greater variability of values in the period of stimulation, which means that this parameter was more unstable, probably caused by the stimulation with the recording of the unknown voice.

The values of diastolic and mean blood pressure increased during the pre-stimulation and stimulation periods and decreased when the stimulation message ended, returning almost to the pre-stimulation values. These results are consistent with the variation of pulse frequency values, because these vital parameters are physiologically related.

Table 5.Statistical analysis of the parameters monitored during the periods of pre-stimulation, stimulation and post-stimulation with a familiar voice (p-values of the various tests)

	Friedman test			ANOVA
Parameters	p-values of the minimum value	p-values of the maximum value	p-values of the standard deviation	p-values of the mean
Pulse frequency	0.881	0.850	0.458	0.914
Systolic blood pressure	0.135	0.601	0.974	0.120
Diastolic blood pressure	0.763	**0.048**	0.808	**0.048**
Mean blood pressure	0.601	0.368	0.601	0.101
Peripheral body temperature	**0.001**	**0.002**	0.222	**0.023**
Oxygen saturation level	0.754	0.949	0.930	0.292

Bold – Statistically significant (p-value < 0.10).

The results of the Friedman test and ANOVA for the parameters monitored during the periods of pre-stimulation, stimulation and post-stimulation with a familiar voice are described in Table 5. As can be seen, statistically significant differences were identified for the maximum values and means of diastolic blood pressure, and the minimum and maximum values and means in peripheral body temperature.

As mentioned above, Walker et al. (1998) evaluated the effect of familiar voices in patients in coma caused by a Traumatic Brain Injury and found no statistically significant changes for the various vital signs. Data analysis revealed only variability in physiological measures and the pulse was a measure that had a greater variability in relation to other physiological parameters. The results of their (Walker et al., 1998) study agree in part with ours.

Proceeding as before, we applied multiple comparison tests in the cases where significant differences occurred. No statistically significant differences were identified between periods for the maximum values of diastolic blood pressure. The mean diastolic blood pressure was statistically significant when comparing the periods of pre-stimulation and stimulation (p-value = 0.070) and in the periods of pre-stimulation and post-stimulation (p-value = 0.065).

No relevant changes in peripheral body temperature were identified in any of the comparative analyses between the periods of stimulation for the minimum values. There were significant differences for the maximum values of the peripheral body temperature, between the periods of pre-stimulation and post-stimulation (p-value = 0.016) and between the periods of pre-stimulation and post-stimulation (p-value = 0.016).

For the mean values of peripheral body temperature there were significant differences when we compared the values of the period of pre-stimulation with the period of stimulation (p-value = 0.025) and when we compared the values of the periods of pre-stimulation and post-stimulation (p-value=0.032).

There was a clinically relevant increase in the minimum values of systolic, diastolic and mean blood pressure when stimulation began which continued to increase when the stimulus message ended. This increase was unexpected and had not been previously reported in the literature. We chose to designate this event "separation anxiety episode", which has already been described for children (Bar-Haim, Dan, Eshel, and Sagi-Schwartz, 2007; Mofrad, Abdullah, Samah, Mansor, and Baba, 2009; Verissimo, Alves, Monteiro and Oliveira, 2003). Usually, such episodes are related to expressions of concern, anxiety and fear when a significant person leaves or is absent. Thus, we believe that there was a reaction linked to the termination of the stimulus message, and a generation of a sense of abandonment and absence.

There was an increase in the values of standard deviation of the pulse frequency in the period of pre-stimulation, during the period of stimulation, followed by a decrease between the period of stimulation and the period of post-stimulation. These results indicate that the variability of values was higher in the period of stimulation.

Similarly, clinically relevant variability in the values of systolic, diastolic and mean blood pressure was found. The peripheral body temperature and oxygen saturation levels remained almost unchanged throughout the periods of stimulation.

Next we present in Table 6 the results of the analysis for comparing the unknown with the familiar voice.

Table 6.Comparative statistical analysis of the minimum values, maximum values, standard deviation and mean of the parameters monitored during the periods of stimulation and post-stimulation with an unknown and familiar voice (p-values of the various tests)

	Stimulation period				Post-stimulation period			
	Wilcoxon test			t test	Wilcoxon test			t test
Parameters	p-values of the minimum value	p-values of the maximum value	p-values of the standard deviation	p-values of the mean	p-values of the minimum value	p-values of the maximum value	p-values of the standard deviation	p-values of the mean
Pulse frequency	**0.061**	0.150	0.396	**0.027**	**0.064**	**0.066**	0.357	**0.033**
Systolic blood pressure	0.357	0.423	0.246	0.348	0.285	0.216	**0.010**	0.446
Diastolic blood pressure	0.348	0.461	0.271	0.401	0.138	**0.040**	**0.002**	0.308
Mean blood pressure	0.275	0.461	0.210	0.384	0.278	**0.097**	**0.003**	0.472
Peripheral body temperature	**0.073**	**0.047**	0.250	**0.038**	**0.027**	**0.043**	0.750	**0.025**
Oxygen saturation level	0.215	**0.053**	0.234	0.127	0.148	0.219	0.438	0.197

Bold – Statistically significant (p-value < 0.10).

As can be seen in the Table 6 there are significant difference in some of the parameters both during the stimulation and the post-stimulation periods. This indicates that the comatose patients reacted differently depending on the type of voice.

When we compared the responses with an unknown voice and those with familiar voices we found that, for all parameters, there were more significant differences post-stimulation than during the period of stimulation. The events that we designated as "separation anxiety episodes" occur more frequently when the voices of family end than for those of unfamiliar people.

A clinically relevant difference in standard deviation values between the stimulation with an unknown voice and a familiar voice for the transition between the period of stimulation and the period of post-stimulation was observed. There was a tendency for participants to normalise the values of the monitored parameters whilst stimulated with an unknown voice and therefore the majority of the participants showed a decrease of the standard deviation values. When stimulated with a familiar voice, we concluded that there was an increase in the number of participants showing higher standard deviation values after stimulation terminated. This provides additional evidence for the "separation anxiety episodes" described previously.

3.2.2. Statistical and Clinical Analysis in the Phases of the Stimulus Message

Table 7 contains the results of the tests performed to study the period of auditory stimulation with the stimulus message of the unknown voice. These tests compare the values during the transition of the three phases of presentation and orientation, information and functional assessment and stimulation.

Table 7.Statistical analysis of the parameters monitored during the phases of presentation and orientation, information and functional assessment and stimulation with an unknown voice (p-values of the various tests)

	Friedman test			ANOVA
Parameters	p-values of the minimum value	p-values of the maximum value	p-values of the standard deviation	p-values of the mean
Pulse frequency	0.989	0.444	0.740	**0.020**
Systolic blood pressure	0.974	0.974	0.615	0.565
Diastolic blood pressure	0.710	0.830	0.330	0.628
Mean blood pressure	0.974	0.830	0.897	0.549
Peripheral body temperature	0.519	0.556	1.000	0.630
Oxygen saturation level	0.345	0.936	**0.050**	0.300

Bold – Statistically significant (p-value < 0.10).

We found that the Friedman test was significant only for the values of standard deviations of the oxygen saturation level. The ANOVA was significant for the mean of the pulse frequency.

As before, we performed multiple comparisons in both situations where differences were considered significant. The results of the Wilcoxon test for the different paired phases of standard deviation values of oxygen saturation level indicated that no statistically significant differences exist between them. For the mean of the pulse frequency we applied multiple t tests, which detected significant differences between the means of the phase of information and the phase of functional assessment (p-value = 0.066) and between the means of the phase of presentation and orientation and the phase of functional assessment and stimulation (p-value = 0.032). These changes may be related to the contents of the phase of functional assessment and stimulation being more provocative, or because the comatose patient knows that if unresponsive to verbal stimulation, a painful stimulus will be applied. Focusing on the clinical analysis, the minimum value of the pulse frequency did not change significantly. The same happened for the values of the other parameters that remain similar throughout the three phases of the stimulus message, which is consistent with the tests previously presented.

From the clinical analysis of the maximum values of the pulse frequency, we concluded that no changes exist in this parameter between the phase of presentation and orientation and the phase of information. However when we observed the values of the phase of information and the phase of functional assessment and stimulation there was an increase in the values, which means that the participants reacted when they heard an unknown voice asking them to move. We also looked at the range of the maximum values of the systolic blood pressure. The minimum values of this range for the total sample, showed a relevant change between the phase of presentation and orientation and the phase of information. These values move back toward normal ranges in the functional assessment and stimulation phases.

We then clinically examined the minimum and maximum values, means and standard deviations of the means of the total sample. The global observation of the values, indicated that the differences were not significant (except in the minimum values of the mean of the systolic blood pressure that increased between the phase of presentation and orientation and the phase of information) in the various parameters evaluated or in the transition of the three phases of the stimulus message.

Analysing the variations in the values of standard deviations, we found that overall there were no clinically relevant changes in all parameters monitored when comparing the values of the phase of presentation and orientation and the phase of information. If we compare the values of the phase of information with the values of the phase of functional assessment and stimulation, we observed an increase in standard deviation values (and therefore variability) in pulse frequency, systolic blood pressure, diastolic blood pressure and mean blood pressure.

In the overall statistical analysis for the three phases of the familiar stimulus message, we found (see Table 8) that the Friedman test was significant for the maximum values of the diastolic blood pressure, minimum and maximum values of the peripheral body temperature and maximum values and standard deviation of the oxygen saturation level. ANOVA was significant for the mean values of peripheral body temperature.

As before, we have carried out multiple comparisons. We found no statistically significant differences in most of the cases.

We only found statistically significant differences in the maximum values of oxygen saturation level between the phases of information and functional assessment and stimulation (p-value = 0.031) and in the mean values of peripheral body temperature between the phase of presentation and orientation and the phase of functional assessment and stimulation (p-value = 0.037).

Table 8.Statistical analysis of the parameters monitored during the phases of presentation and orientation, information and functional assessment and stimulation with a familiar voice (p-values of the various tests)

	Friedman test			ANOVA
Parameters	p-values of the minimum value	p-values of the maximum value	p-values of the standard deviation	p-values of the mean
Pulse frequency	0.516	0.856	0.156	0.430
Systolic blood pressure	0.710	0.437	0.135	0.331
Diastolic blood pressure	0.491	**0.092**	0.114	0.261
Mean blood pressure	0.601	0.135	0.176	0.271
Peripheral body temperature	**0.074**	**0.082**	1.000	**0.040**
Oxygen saturation level	0.938	**0.094**	**0.045**	0.894

Bold – Statistically significant (p-value < 0.10).

We also analysed each of the variables to verify if any clinically relevant changes existed in the transition of the various phases of the stimulus message with a familiar voice. The standard deviations for the different pulse frequency values remained relatively stable in the transition between the phase of presentation and orientation and the phase of information, but there was a large increase in the values when we transited to the phase of functional assessment and stimulation. In the values of systolic, diastolic and mean blood pressure we observed a decrease in standard deviation values when we transit from the phase of presentation and orientation to the phase of information, which increased again when we transit to the phase of functional assessment and stimulation. The values for peripheral body temperature remained relatively unchanged during the three phases of stimulation and the values of oxygen saturation level followed the same line of changes described above for the pressures.

Table 9 contains the results of the tests performed in order to assess differences between the responses with the unknown and the familiar voices.

We can observe that during the phase of presentation and orientation there were statistically significant differences in some of the variables in all the phases for the pulse frequency. The other parameters exhibited fewer significant differences with rather high p-values.

Table 9. Comparative statistical analysis of the minimum values, maximum values, standard deviation and mean of the parameters monitored during the phases of presentation and orientation, information and functional assessment and stimulation with an unknown and familiar voice (p-values of the various tests)

	Presentation and orientation				Information				Functional assessment and stimulation			
	Wilcoxon test			t test	Wilcoxon test			t test	Wilcoxon test			t test
Parameters	p-values of the minimum value	p-values of the maximum value	p-values of the standard deviation	p-values of the mean	p-values of the minimum value	p-values of the maximum value	p-values of the standard deviation	p-values of the mean	p-values of the minimum value	p-values of the maximum value	p-values of the standard deviation	p-values of the mean
Pulse frequency	0.063	0.004	0.604	0.047	0.402	0.021	0.887	0.079	0.328	0.434	1.000	0.070
Systolic blood pressure	0.846	0.910	0.078	0.778	0.625	0.695	0.719	0.496	0.922	0.625	0.322	0.901
Diastolic blood pressure	0.574	0.264	0.049	0.692	0.846	0.922	0.826	0.873	0.643	0.922	0.668	0.649
Mean blood pressure	0.232	0.770	0.047	0.791	0.359	0.826	0.859	0.512	0.570	0.826	0.643	0.980
Peripheral body temperature	0.145	0.127	1.000	0.125	0.141	0.125	1.000	0.076	0.090	0.094	1.000	0.051
Oxygen saturation level	0.063	0.523	0.945	0.213	0.633	0.066	0.500	0.283	0.734	0.461	0.906	0.247

Bold – Statistically significant (p-value < 0.10).

3.3. Statistical and Clinical Analysis of the Monitored Curves

3.3.1. Statistical and Clinical Analysis in the Periods of Stimulation

The curve analysis proved important because it allowed us to corroborate the results obtained previously for the parameters and allowed a more rigorous study of the influence of auditory stimulation, in the sense that we worked directly with physiological variables monitored but not processed by the multiparameter monitor. It was also important to analyse the curves, because we worked with a sampling frequency much higher than the sampling frequency of the parameters.

Table 10 shows that there were no statistically significant differences in any of the monitored curves with an unknown voice.

We concluded that the most clinically relevant changes occurred in the means of the curve of arterial pressure. The standard deviations showed a progressive increase over the three periods of stimulation, which means that there was a greater variability in the values found. No relevant clinical changes were found for the curve of pulse plethysmography over the period of analysis.

Table 11 shows that there were no statistically significant differences in the values of the three periods of stimulation with the stimulus message of the familiar voice.

Table 10.Statistical analysis of the curves monitored during the periods of pre-stimulation, stimulation and post-stimulation with an unknown voice (p-values of the various tests)

	Friedman test			ANOVA
Curves	p-values of the minimum value	p-values of the maximum value	p-values of the standard deviation	p-values of the mean
RR intervals	0.726	0.114	0.938	0.157
Arterial pressure	-	-	0.222	0.868
Pulse plethysmography	-	-	0.897	0.304

Table 11.Statistical analysis of the curves monitored during the periods of pre-stimulation, stimulation and post-stimulation with a familiar voice (p-values of the various tests)

	Friedman test			ANOVA
Curves	p-values of the minimum value	p-values of the maximum value	p-values of the standard deviation	p-values of the mean
RR intervals	0.994	0.205	0.852	0.973
Arterial pressure	-	-	0.601	0.105
Pulse plethysmography	-	-	0.947	0.685

The mean values of the curve of arterial pressure showed a clinically relevant increase when we started the stimulation, but also continued to increase when the auditory stimulation ended. For the standard deviation of the curve of pulse plethysmography we observed no changes for the transition between the pre-stimulation and stimulation period, however, when the stimulation ended there was a decrease in all values.

Table 12 contains the results of the tests performed to compare the responses obtained with the unknown and the familiar voices. Unilateral test were selected because it was expected that the comatose patients would have a more exacerbated response during the stimulation with a familiar voice than during the stimulation with an unknown voice.

As can be seen, significant differences were only detected in the RR intervals, both in the stimulation and the post-stimulation periods.

From an overall clinical (relevance) analysis, we concluded that the reaction of participants was more marked on stimulation with a familiar voice than to an unknown voice on every curve and on the two transitions in question.

Table 12. Comparative statistical analysis of the minimum values, maximum values, standard deviation and mean of the curves monitored during the periods of stimulation and post-stimulation with an unknown and familiar voice (p-values of the various tests)

	Stimulation period				Post-stimulation period			
	Wilcoxon test			t test	Wilcoxon test			t test
Curves	p-values of the minimum value	p-values of the maximum value	p-values of the standard deviation	p-values of the mean	p-values of the minimum value	p-values of the maximum value	p-values of the standard deviation	p-values of the mean
RR intervals	0.107	0.201	**0.047**	**0.011**	**0.049**	0.332	**0.070**	**0.017**
Arterial pressure	-	-	0.385	0.379	-	-	0.385	0.443
Pulse plethysmography	-	-	0.271	0.222	-	-	0.287	0.390

Bold – Statistically significant (p-value < 0.10).

3.3.2. Statistical and Clinical Analysis in the Phases of the Stimulus Message

The results of the statistical tests used to study the monitored curves resulting from stimulation with and unknown voice are presented in Table 13.

Table 13. Statistical analysis of the curves monitored during the phases of presentation and orientation, information and functional assessment and stimulation with an unknown voice (p-values of the various tests)

	Friedman test			ANOVA
Curves	p-values of the minimum value	p-values of the maximum value	p-values of the standard deviation	p-values of the mean
RR intervals	0.519	0.586	0.737	**0.047**
Arterial pressure	-	-	0.974	0.588
Pulse plethysmography	-	-	0.285	**0.097**

Bold – Statistically significant (p-value < 0.10).

We found significant differences for the means of the RR intervals and the means of the curve of pulse plethysmography. Thus, in these cases we applied multiple comparison tests. We only found significant differences (p-value = 0.037) in the case of the RR intervals between the phase of presentation and orientation, and the phase of functional assessment and stimulation.

From a clinical point of view we found that the mean of the minimum values of the RR interval decreased considerably during the phase of functional assessment and stimulation. However, the values of standard deviation increased. This increase in the values of standard deviation means that the entire sample exhibits a greater variability in the phase of functional assessment and stimulation. The decrease in the other values means that, in this phase, there has been an increase in heart rate and consequently in the pulse frequency.

For the mean values of the curve of arterial pressure, the standard deviation value of the phase of functional assessment and stimulation was higher than the others, which reveals that, although the mean values are not relevantly different, there is a greater variability among the participants in the study in this phase. In the curve of pulse plethysmography, we concluded that higher values of standard deviation were recorded in the information phase. Relative to the stimulation with a familiar voice we can see in Table 14, that there were no statistically significant differences in any of the cases under analysis.

Table 14.Statistical analysis of the curves monitored during the phases of presentation and orientation, information and functional assessment and stimulation with a familiar voice (p-values of the various tests)

	Friedman test			ANOVA
Curves	p-values of the minimum value	p-values of the maximum value	p-values of the standard deviation	p-values of the mean
RR intervals	0.993	0.774	0.177	0.428
Arterial pressure	-	-	0.763	0.236
Pulse plethysmography	-	-	0.682	0.766

Table 15. Comparative statistical analysis of the minimum values, maximum values, standard deviation and mean of the curves monitored during the phases of presentation and orientation, information and functional assessment and stimulation with an unknown and familiar voice (p-values of the various tests)

	Presentation and orientation				Information				Functional assessment and stimulation			
	Wilcoxon test			t test	Wilcoxon test			t test	Wilcoxon test			t test
Curves	p-values of the minimum value	p-values of the maximum value	p-values of the standard deviation	p-values of the mean	p-values of the minimum value	p-values of the maximum value	p-values of the standard deviation	p-values of the mean	p-values of the minimum value	p-values of the maximum value	p-values of the standard deviation	p-values of the mean
RR intervals	0.016	0.389	0.438	0.014	0.027	0.344	0.531	0.036	0.250	0.754	0.094	0.056
Arterial pressure	-	-	0.695	0.760	-	-	0.557	0.504	-	-	1.000	0.963
Pulse plethysmography	-	-	0.674	0.796	-	-	0.557	0.411	-	-	0.311	0.427

Bold – Statistically significant (p-value < 0.10).

From a clinical perspective, we found that in the minimum values of RR intervals, the most significant alteration occurred during the phase of functional assessment and stimulation, registering a slight decrease in all the values. For the means of the RR interval the phase of functional assessment and stimulation registered the lowest values indicating that it is in this phase where the increased values of the pulse frequency occurred.

Table 15 shows the results of comparing the values obtained in the analysis of the curves monitored during the three phases of the stimulus message with an unknown voice and a familiar voice. In this analysis we chose to apply two-tailed tests because it was not possible to predict that if there was a reaction from the comatose patients at various phases of stimulation, this would be exacerbated in the case of stimulation with an unknown voice or a familiar voice.

As can be seen in Table 15, only the RR-intervals show significant differences. They occur in all phases but only in some of the variables.

Conclusions

It is considered good practice to communicate with comatose patients because they may be able to hear and some patients may benefit from verbal stimulation which may assist in reducing stress, helping patients preserve self-identity and self-esteem, and provide orientation, meaningful sensory input and reduce social isolation.

This study investigated the effects of recorded voice messages provided by an unknown person and by a significant person on key measures of physiological functioning in comatose patients in an ICU.

Results showed that the stimulation with an unknown and familiar voice generated an increase in heart rate values and in the values of arterial pressure. This was corroborated by the analysis of monitored curves. However, for a familiar voice the values of these parameters continued to increase post-stimulation, which revealed that participants developed episodes of separation anxiety. No changes in body temperature occurred during the stimulation.

When the phases of the stimulus message were analysed, during the stimulation with an unknown voice, the functional assessment and stimulation phase showed a larger variability in the values of heart rate, arterial pressure and oxygen saturation level. This was corroborated by the analysis of the curves. Stimulation with a familiar voice resulted in greater changes for the same parameters during the functional assessment and stimulation phase and participants also presented relevant change during the presentation and orientation phases.

Acknowledgments

The authors would like to thank AT, AF, AS, FF, JF, LB, MA, MC, NP, RG and their families, André Baldaia, Aníbal Marinho, Fernando Rua, José Pinho, Sara Demain and Rui Alcovia.

This work was partially funded by FEDER through the Operational Program Competitiveness Factors - COMPETE and by National Funds through FCT - Foundation for Science and Technology in the context of the project FCOMP-01-0124-FEDER-022682 (FCT reference PEst-C/EEI/UI0127/2011) and the project FCOMP-01-0124-FEDER-022690 (FCT reference PEst-C/MAT/UI4106/2011).

References

Adit (2006).Sociogram Software. Dwyran: Adit.

Alasad, J. and Ahmad, M. (2005) Communication with critically ill patients. *Journal of Advanced Nursing* 50(4), 356-362.

Allen, J. (2007). Photopletysmography and its application in clinical physiological measurement. *Physiological Measurement 28*, R1-R39.

Anderson, R. (2007). *Thematic content analysis (TCA): descriptive presentation of qualitative data*. Palo Alto: Institute of Transpersonal Psychology.

Ashworth, P. (1987). The needs of the critically ill patient. *Intensive Care Nursing* 3(4), 182-190.

Baker, C. and Melby, V. (1996). An investigation into the attitudes and practices of intensive care nurses towards verbal communication with unconscious patients. *Journal of Clinical Nursing* 5, 185-192.

Bar-Haim, Y., Dan, O., Eshel, Y., and Sagi-Schwartz, A. (2007). Predicting children's anxiety from early attachment relationships. *Journal of Anxiety Disorders* 21, 1061-1068.

Bastin, G. (1980). Les Techniques Sociometriques. Paris: Seli Arsian.

Bland, W., Busam, V., Gunlogson, B., Mekkes, G., and Saunders, A. (2006). Audacity (Version 1.2.6). Source Forge.

Bolander, V. (1994). *Sorensen and Luckmann's Basic Nursing: A Psychophysiologic Approach* (3rd ed.). Philadelphia: W. B. Saunders.

Boly, M., Garrido, M., Gosseries, O., Bruno, M., Boveroux, P., Schnakers, C., Massimini, M., Litvak, V., Laureys, S. and Friston, K. (2011). Preserved feedforward but impaired top-down processes in the vegetative state. *Science* 332(5), 858-862.

Castells, F., Laguna, P., Sornmo, L., Bollmann, A., and Roig, J. M. (2007). Principal component analysis in ECG signal processing. *EURASIP Journal on Advances in Signal Processing* 74580, 1-21.

Clifford, G., Azuaje, F., and McSharry, P. (2006). *Advanced Methods and Tools for ECG Data Analysis*. Norwood: Artech House.

Datex-Ohmeda. (2003). *Datex-Ohmeda S/5 Collect - User's Reference manual*. Helsinki: Datex-Ohmeda.

Doman, G., Wilkinson, R., Dimancescu, M. and Pelligra, R. (1993).The effect of intensive multi-sensory stimulation on coma arousal and recovery. *Neuropsychological Rehabilitation* 3(2), 203-212.

Elliot, R. and Wright, L. (1999). Verbal communication: what do critical care nurses say to their unconscious or sedated patients? *Journal of Advanced Nursing* 29(6), 1412-1420.

Fischer, C. (2006). *Qualitative research methods for psychologists: introduction through empirical studies*. New York: Academic Press.

Holeckova, I., Fischer, C., Giard, M., Delpuech, C. and Morlet, D. (2006). Brain responses to a subject's own name uttered by a familiar voice. *Brain Research* 1082, 142-152.

Hoonakker, P., Carayon, P., Douglas, S., Schultz, K., Walker, J. and Wetterneck, T. (2008). Communication in Intensive Care Units and the Relation with Quality of Care and Patient Safety from a Nurse Perspective. In L. Sznelwar, F. Mascia and U. Montedo (Eds.), *Human Factors in Organizational Design and Management – IX* (pp. 715-721). Santa Monica: IEA Press.

Hoyt, J. (1996). A gentle approach: interacting with a person who is semi-conscious or presumed in coma. *Issues in Law and Medicine* 12(1), 71-78.

Jennett, B. (2002). The Glasgow Coma Scale: History and current practice. *Trauma* 4, 91-103.

Johnson, S., Omery, A. and Nikas, D. (1989). Effects of conversation on intracranial pressure in comatose patients. *Neurologic Aspects of Critical Care* 18(1), 56-63.

Jones, R., Hux, K., Morton-Anderson, A. and Knepper, L. (1994). Auditory stimulation effect on a comatose survivor of traumatic brain injury. *Archives of Physical Medicine Rehabilitation* 75, 164-171.

Kerlinger, F. and Lee, H. (2000). *Foundations of behavioral research* (4th ed.). Belmont: Wadsworth - Thomson Learning.

LaPuma, J., Schiedermayer, D., Galias, A. and Siegler, M. (1990). Talking to comatose patients. *Neurotrauma Medical Report* 4, 1-3.

McSharry, P. and Clifford, G. (2006). Models for ECG and RR Interval Processes. In C. Gari D, F. Azuajeand P. E. McSharry (Eds.), *Advanced Methods and Tools for ECG Data Analysis* (pp. 101-133). Norwood: Artech House.

Mofrad, S., Abdullah, R., Samah, B., Mansor, M. and Baba, M. (2009). Maternal psychological distress and separation anxiety disorders in children. *European Journal of Social Sciences* 8(3), 386-394.

Murteira, B., Ribeiro, C., Silva, J. and Pimenta, C. (2007). *Introdução à Estatística* (2nd ed.). Lisboa: McGraw-Hill.

Oliveira, J., Maojo, V., Martin-Sanchez, F. and Pereira, A. (2005). *Biological and Medical Data Analysis*. Berlin: Springer.

Perrin, F., Schnakers, C., Schabus, M., Degueldre, C., Goldman, S., Brédart, S., Faymonville, M., Lamy, M., Moonen, G., Luxen, A., Maquet, P. and Laureys, S. (2006). Brain Response to one's own name in vegetative state, minimally conscious state, and locked-in syndrome. *Archives of Neurology* 63, 562-569.

Prados, J. (2000). *Sociologia de Los Grupos Escolares: Sociometría y Dinámica de Grupos*. Almería: Universidad de Almería.

Puggina, A. (2006) *O uso da música e de estímulos vocais em pacientes em estado de coma: relação entre estímulo auditivo, sinais vitais, expressão facial e escalas de Glasgow e Ramsay*. M.Sc. Thesis, Universidade de São Paulo, São Paulo, Brasil.

Rangayyan, R. (2002). Biomedical Signal Analysis - A case-Study Approach. New York: Wiley-Interscience.

Reisner, A., Clifford, G., and Mark, R. (2006). The Physiological Basis of the Electrocardiogram. In G. Clifford, F. Azuage and P. Mcsharry (Eds.), *Advanced Methods and Tools for ECG Data Analysis* (pp. 1-25). Norwood: Artech House.

Ritchie, J., and Spencer, L. (1994). Qualitative data analysis for applied policy research. In A. Brymanand and R. Burgess (Eds.), *Analysing Qualitative Data* (pp. 173-194). London: Routledge.

Saritha, C., Sukanya, V., and Murthy, Y. N. (2008). ECG Signal Analysis Using Wavelet Transforms. *Bulg. J. Phys. 35*, 68-77.

Simões, J., Jesus, L. and Voegeli, D. (2012).Verbal communication with unconscious patients: Content analysis and development of a stimulus message. *Journal of Advanced Nursing*, Submitted.

Simões. J., Jesus, L., Voegeli, D., Couto, P., Fernandes, J. and Morgado, M. (2011). Assessment of comatose patients: a Portuguese instrument based on the Coma Recovery Scale – Revised and using nursing standard terminology. *Journal of Advanced Nursing* 67(5), 1129-1141.

Simoliuniene, R., Krisciukaitis, A., Macas, A., Baksyte, G., Saferis, V., and Zaliunas, R. (2008). Principal Component Analysis Based Method for Detection and Evaluation of ECG T-Wave Alternans. *Computers in Cardiology* 35, 757-760.

Stanhope, M., and Lancaster, J. (2011). *Public Health Nursing: Population-Centered Health Care in the Community (8th ed.)*. St. Louis: Mosby.

Turnock, C. (1991). Communicating with patients in ICU. *Nursing Standard* 9(5), 38-40.

Urden, L., Stacy, K., and Lough, M. (2010). *Critical Care Nursing: Diagnosis and Management (6th ed)*. St. Louis: Mosby.

Veríssimo, M., Alves, S., Monteiro, L., and Oliveira, C. (2003). Ansiedade de separação materna e adaptação psicossocial ao pré-escolar. *Psicologia, Saúde e Doenças* 4(2), 221-229.

Verity, S. (1996). Communicating with sedated ventilated patients in intensive care: focusing on the use of touch. *Intensive and Critical Care Nursing* 12, 354-358.

Walker, J., Eakes G. and Siebelink E. (1998). The effects of familial voice interventions on comatose head-injured patients. *Journal of Trauma Nursing* 5 (2), 41-45.

In: Comas and Syncope: Causes, Prevention and Treatment ISBN 978-1-62100-603-9
Editors: Eduardo Silva and Gabriel Cruz.

Chapter 2

SUDDEN CARDIAC DEATH IN PEDIATRICS: GENETIC BASIS AND CLINICAL TREATMENT

***Georgia Sarquella-Brugada*[1]*, Oscar Campuzano*[2*]*, Paola Berne*[3]*, Josep Brugada*[3]*, Pedro Brugada*[4] *and Ramon Brugada*[2]**

[1]Unit of Arrhythmias, Hospital Sant Joan de Deu, University of Barcelona, (Spain)
[2]Cardiovascular Genetics Center, University of Girona–IdIBGi, Girona (Spain)
[3]Unit of Arrhythmias, Hospital Clinic de Barcelona,University of Barcelona, (Spain)
[4]Heart Rhythm Management Center, Cardiovascular Division,UZ Brussel-Vrije Universiteit Brussel, Brussels, (Belgium)

ABSTRACT

In last 15 years, cardiology has experienced an amazing progress in genetics and molecular biology. These advances help to develop new methods of prevention, diagnosis and clinical treatment of cardiac pathologies. The majority of these diseases have an incomplete penetrance; therefore patients may be unaware of their illness. Sudden cardiac death (SCD) is a rare but socially devastating event. In several cases, physical activity can be the trigger for SCD as first symptom. Most common causes of SCD are congenital electrical disorders and structural heart diseases although a significant number of SCD remain unexplained. In the pediatric population, in up to 50% of SCD cases, sudden death (SD) is the first and only clinical manifestation of an inherited cardiac disease that had remained undetected by conventional clinical investigations. Concretely, sudden infant death syndrome (SIDS) accounts for 10% of infant mortality in the first year of life. Most of the SCD in pediatric population are due to channelopathies and cardiomyopathies. This review focuses on recent advances in genetic channelopathies and cardiomyopathies in pediatric population and their clinical treatment.

* Corresponding Author: Oscar Campuzano Larrea, PhD Cardiovascular Genetics Center Parc Cientific i Tecnologic, Pic Peguera 11 Girona 17003, Spain, Email: oscar@brugada.org

INTRODUCTION

Sudden death syndrome (SDS) in children is defined as the natural and unexpected event that occurs in an apparently healthy infant or young child or whose disease was not severe enough to predict a fatal outcome, and in which a thorough postmortem examination fails to demonstrate an adequate cause of death [1]. Despite these attempts, the definitions and protocols used for diagnosing SDS still remains a diagnosis of exclusion [2].

SDS affects pediatric population (fetus, newborns, children and adolescents), with an individual risk of 1-3 per 100,000 person-year [3-5]. It accounts for 10% of infant mortality in the first year of life [1, 2, 6-9]. The number of quality-adjusted years of life lost is considerable, as most of these victims would have been expected to have a normal life expectancy with a minimum of symptoms with the optimal treatment.

SDS is a disorder with many mechanisms resulting in or predisposing to its development, including infections and several genetic abnormalities [10]. In the last 20 years, the advanced in genetics of SD has been tremendous and it has been identified hundreds of mutations in several genes encoding cardiac proteins. The majority of SCD in children and young people are due to hereditary and congenital cardiac channelopathies, cardiomyopathies, coronary artery anomalies and/or aortic root dissection [11, 12].

There are various elements necessary to achieve coordinated cardiac activity responsible for transmitting electrical and mechanical impulse through the cardiac myocytes, including ionic currents, ion channels and structural proteins. The complexity of this process remains the major limitation for understanding arrhythmogenesis [13]. Functional analysis of ion channels has enabled a better understanding of basic arrhythmogenic mechanisms but it was not until the development of genetics and the discovery of disease causing mutations family when it has been extrapolated from basic science to clinical practice [14].

CHANNELOPATHIES

Channelopathies are heart diseases induced by mutations in genes encoding cardiac ion channels. It is not associated with structural cardiac abnormalities and its first manifestation may be SCD. Moreover, some of these diseases are not accompanied by changes in the electrocardiogram (ECG), which makes it more difficult diagnosis. Given that these diseases are determined by a genetic defect, it is hoped that genetic testing can contribute substantially to the diagnosis, prevention and treatment. In 1976 was published the first association between SIDS and a cardiac disorder, the long QT syndrome (LQTS) [15, 16]. Since this association, genetic and/or clinical correlations between channelopathies and SIDS have been found in several studies [17]. To date, up to 35% of cases of SCD in young people may be caused by a genetic mutation in ion channels [18].

Long QT Syndrome

LQTS is a clinically and genetically heterogeneous cardiac channelopathy with a prevalence of 1 in 2500 individuals. It is characterized for a prolongation of the QT interval

(QTc>480ms). The clinical presentation can be variable, ranging from asymptomatic patients to episodes of syncope and SCD due to ventricular tachyarrhythmias (*torsade de pointes*) in the setting of a structurally normal heart in otherwise healthy individual [19, 20]. Due to the ability to identify the individuals at risk through ECG analysis, massive population screening by ECG has been performed in certain regions, with success of lowering rates of SD among infants and athletes [21-23]. LQTS is a major cause of SCD among young people [17], and recent progress in molecular biology has clarified that LQTS partly contributes to SIDS [24, 25].

LQTS can be congenital or acquired: the latter being a consequence of QT prolonging drugs and less frequently heart block and myocardial infarction (MI) [26]. The congenital form is associated with mutations in ion channels and/or associated proteins. Inheritance of LQTS can follow an autosomal dominant (Romano-Ward syndrome) or recessive (Jervell and Lange-Nielsen syndrome) transmission pattern. To date, more than 600 mutations and splice-site altering mutations have been identified in 14 LQTS-susceptibility or LQTS overlap-susceptibility genes (Table1) [27].

Approximately a 75% of clinically definite LQTS are caused by mutations in 3 genes: *kcnq1* (LQT1), *kcnh2* (LQT2), and *scn5a* (LQT3). The remaining 25% have been identified in a variety of ion channels or channel-interacting proteins [28]. The most common form, LQTS type 1, caused by mutations in *kcnq1* is responsible for 40-50% of the cases of prolonged QT interval [29, 30]. It was also described the gene *kcnh2* (*HERG, human-ether-a-go-go-related*) that codifies the α-subunit of Ikr. The mutations in this gene suppose a 35%-45% of cases (LQTS type 2) [31, 32]. The β-subunit is codified by *kcne2* (MiRP1 protein); the mutations in *kcne2* gene induce LQTS type 6, a rare type (<1%) that also induce a loss of function. Other gene is *kcne1*, responsible for 2%-5% of cases (LQTS type 5), which can alter both iks and ikr [33]. The *kcnj2* gene is also implicated in LQTS. It codifies Kir2.1 protein; the mutations are associated to loss of function (LQTS type 7 or Tawil-Anderson syndrome). The incidence is very low and infrequently is associated to SCD [34]. Recently, is has been associated *kcnj5* gene (Kir3.4 –also named GIRK4-) to LQTS (LQTS type 13). The mutation induces a loss of function [35]. Mutations in the LQTS give rise especially to a gain of function in the sodium current (inappropriate prolonged entry of Na^+ into the myocyte), by mutations in *scn5a (*LQTS type 3*)*, the third gene most prevalent in LQTS [36, 37]. Mutations in this gene induce gain of function. The LQTS type 10 is caused by mutations in *scn4b* that codifies for β-subunit of sodium cannel (NaVβ4). The β-subunit plays a key role both in the kinetic regulation and in the a-subunit expression of the sodium channel [38]. In 2008, mutations in *snta-1* gene were associated to LQT (type12) thought a gain of function of the fast sodium channel [39, 40]. It encodes alpha1-syntrophin protein. The LQTS type 9 is caused by mutations in *Cav3* gene [41]. Mutations in this gene induce gain of function of sodium channels, similar as LQTS type3. Calcium channels are also associated to LQTS. Type 8 LQTS (Timothy syndrome*)* has been described with a mutation in the *cacna1c* gene that encodes the pore (Cav1.2) of the L-type cardiac calcium channel. This type of LQTS is uncommon, but it has the highest associated mortality. The mutation induces an enhanced function with Ica abnormality, and loss of the channel dependent voltage, leading to a prolongation of the action potential [42, 43].

Table 1. Ion Channel Diseases

CHANNEL	DISEASE	INHERITANCE	LOCUS	GENE	PROTEIN
Sodium	LQT 3 (RW)	AD	3p21-24	*SCN5A*	Nav1.5
	LQT 10 (RW)	AD	11q23.3	*SCN4B*	Navβ4
	BrS 1	AD	3p21-p24	*SCN5A*	Nav1.5
	BrS 2	AD	3p22.3	*GPD1-L*	glycerol-3-P-DH-1
	BrS 5	AD	19q13.1	*SCN1B*	Navβ1
	BrS 7	AD	11q24.1	*SCN3B*	Navβ3
	Lev-Lenègre Syndrome (PCCD)	AD	3p21	*SCN5A*	Nav1.5
	AF	AD	3p21	*SCN5A*	Nav1.5
		AD	19q13.1	*SCN1B*	Navβ1
		AD	11q23.3	*SCN2B*	Navβ2
		AD	11q24.1	*SCN3B*	Navβ3
	IVF	AD	3p21-24	*SCN5A*	Nav1.5
Sodium related	LQT 9 (RW)	AD	3p25	*CAV3*	M-Caveolin
	LQT 12 (RW)	AD	20q11.2	*SNTA1*	α-Syntrophin
	BrS 10	AD	17p13.1	*MOG1*	RAN-G-release factor
Potassium	LQT 1 (RW)	AD	11p15.5	*KCNQ1*	Kv7.1
	LQT 2 (RW)	AD	7q35-q36	*KCNH2*	hERG Kv11.1
	LQT 5 (RW)	AD	21p22.1-p22.2	*KCNE1*	MinK
	LQT 6 (RW)	AD	21p22.1-p22.2	*KCNE2*	MiRP1
	Anderson Syndrome (LQT 7)	AD	17q23.1-q24.2	*KCNJ2*	Kv2.1 Kir2.1
	LQT 13 (RW)	AD	11q24.3	*KCNJ5*	Kv3.1 Kir3.4
	LQT 1 (JLN1)	AR	11p15.5	*KCNQ1*	Kv7.1
	LQT 5 (JLN2)	AR	21q22.1	*KCNE1*	MinK

CHANNEL	DISEASE	INHERITANCE	LOCUS	*GENE*	PROTEIN
Potassium					
	SQT 1	AD	7q35	*KCNH2*	hERG Kv11.1
	SQT 2	AD	11p15.5	*KCNQ1*	Kv7.1
	SQT 3	AD	17q23	*KCNJ2*	Kv2.1 Kir2.1
		AD	10q22	?	-
		AD	6q14-q16	?	-
		AD	10p11-q21	?	-
		AD	5p15	?	-
	AF	AD	11p15.5	*KCNQ1*	Kv7.1
		AD	12p13	*KCNA5*	Kv3.4 Kir3.4
		AD	21q22	*KCNE2*	MiRP1
		AD	17q23	*KCNJ2*	Kv2.1 Kir2.1
	AF and BrS 6	AD	11q13-q14	*KCNE3*	MiRP2
	AF and SQT	AD	7q35	*KCNH2*	hERG Kv11.1
	BrS 8	AD	12p12.1	*KCNJ8*	Kv6.1 Kir6.1
	CPVT 3	AD	17q23	*KCNJ2*	Kv2.1 Kir2.1
	BrS 9	AD	15q24.1	*HCN4*	hyperpolarization cyclic nucleotide-gated 4
	BrS 11	Sex-linked	Xq22.3	*KCNE5 (KCNE1L)*	potassium voltage-gated channel subfamily E member1 like
	BrS 12	AD	1p13.2	*KCND3*	Kv4.3 Kir4.3
	IVF	AD	12p12.1	*KCNJ8*	Kv6.1 Kir6.1
Potassium related	LQT 11 (RW)	AD	7q21-q22	*AKAP9*	Yotiao
	IVF	AD	7q36.2	*DPP6*	dipeptidyl-peptidase 6
Calcium	BrS 3 and shorter QT (SQT 4)	AD	2p13.3	*CACNA1C*	Cav1.2
	BrS 4 and shorter QT (SQT 5)	AD	10p12.33	*CACNB2B*	voltage-dependent β-2
	SQT 6	AD	7q21-q22	*CACNA2D1*	voltage-dependent α2/δ1
	Timothy Syndrome (LQT 8)	AD	12p13.3	*CACNA1C*	Cav1.2
	LQT 14 (RW)	AD	1q42.1-q43	*RYR2*	Ryanodine Receptor 2
	CPVT 1	AD	1q42.1-q43	*RYR2*	Ryanodine Receptor 2
	CPVT 2	AR	1p13.3	*CASQ2*	calsequestrin 2
	PFHB type I	AD	19q13.33	*TRPM4*	transient receptor potential M4
Calcium related	LQT 4 (RW)	AD	4q25-q27	*ANK2*	Ank-B

AD, Autosomic Dominant; AR, Autosomic Recessive; AF, Atrial Fibrillation; BrS, Brugada Syndrome; CPVT, Catecholaminergic Polymorphic Ventricular Tachycardia; LQT, Long QT; SQT, Short QT; PCCD, Progressive Cardiac Conduction Disease; PFHB, Progressive Familial Heart Block; IVF, Idiopathic ventricular fibrillation; JLN, Jervell and Large-Nielsen Syndrome; RW, Romano-Ward.

This gives rise to an ECG with an extremely long QT interval. Recently, a long QT interval was also associated with a patient that showed a mutation in the cardiac ryanodine receptor gene *ryr2* [44]; however, further studies were required in order to clarify this case report. The type 11 LQT is a disorder of the QT interval caused by a mutation (S1570L) in *akap9* gene, which encodes the protein kinase-A anchor protein-9 (chromosome 7q21-q22). The severity of this type of QT may vary; the most common symptoms are angina, partial or total loss of consciousness and in some cases the SCD [45]. There are other genes as *ank2* (chromosome 4, 4q25-27), which is involved in type 4 LQT syndrome. Although not specific to a channel is included in the group of channelopathies. This gene encodes the protein ankyrin-B which is to adapt different structures in the cell membrane as the Na/K ATPase, Na/Ca and inositol triphosphate receptor. A decrease in the role of ankyrin-B alters calcium homeostasis prolonging repolarization and fatal ventricular arrhythmias generated [46-48]. The syntrophins are cytoplasmic proteins that are part of the protein complex associated with dystrophin.

How to Follow a Child with Long QT Syndrome?

In all patients, adults [49] and children [50], beta-blocker administration at high doses is highly recommended. The dose is adjusted according to the medical tolerance to these drugs [51]. However, controversy exists regarding the efficacy of cardioselective beta-blockers such as atenolol [52-56].

Despite beta-blockade, some patients may remain at risk as they persist symptomatic. Different strategies have been proposed: despite defibrillator implantation, some may consider left cardiac sympathetic neural denervation [57]. Defibrillator implantation is mandatory for those patients having had an aborted SD and for those at risk of fatal arrhythmias [58]. Avoidance of QT prolonging drugs is crucial for these patients (www.torsades.org). Water activities should be contraindicated in all types of LQTS, as high intensity exercise [59].

Study of the Relatives of a Patient with LQT

In patients with LQT is important to have a good family history, and do a baseline ECG to all family members (parents, children, siblings).

Data on the clinical presentation and genotype-phenotype correlation of patients with congenital LQTS diagnosed at perinatal through infantile period are limited [60-62]. A nationwide survey was conducted to characterize how LQTS detected during those periods is different from that in childhood or adolescence. Patients with LQTS who showed life-threatening arrhythmias at perinatal periods were mostly those with LQT2, LQT3, or no known mutations. Independent of the genotype, aggressive intervention resulted in effective suppression of arrhythmias [63].

A prolonged QT interval corrected for heart rate (QTc) is a major risk factor in patients with LQTS. However, heart rate-related risk in this genetic disorder differs among genotypes. Therefore, risk stratification for life-threatening cardiac events in LQTS patients can be improved by incorporating genotype-specific QT correction for heart rate [64].

It is advisable that the genetic study of these patients. In the case of a known mutation, family members should be studied to rule out that they are carriers [59].

What Should We Do with Asymptomatic Gene Carriers?

The gene carrier should avoid competitive sports and should be treated with beta blockers (level of evidence IIa) because use of beta-blockers decreases the risk of SCD [51] although do not provide full protection.

Short QT Syndrome

SQTS—first described in 2000 [65] — is a highly malignant condition characterized by a short QT interval (<330 ms), with a high sharp T wave and a short interval between the peak and the end of the T wave, leading some clinical manifestations from lack of symptoms to recurrent syncope, and high risk of SCD [66-68] because of ventricular arrhythmias, but atrial fibrillation (AF) is also commonly seen in patients with SQTS [69, 70]. Clinical manifestations may appear as early as childhood, and so it is considered a possible cause of SD in nursing infants [17, 71, 72].

The genetic origin of this condition has been reported recently, with an autosomal dominant pattern of transmission and a high penetrance (Table1). The mutations that induce this syndrome are located on 6 genes, of which 3 (*kcnq1*, *kcnj2*, and *kcnh2*) encode potassium channels, with enhanced function and, therefore, shortened repolarization [71]. The origin of this entity has been recently described, with autosomal dominant inheritance pattern and high penetrance. The burnout syndrome type 1 has been associated with two mutations in *kcnh2* (hERG protein) that induce a rapid activation of potassium currents, with a gain of function of IKr and a shortening of ventricular action potentials. Generally, cardiac events are associated with adrenergic in situations such as noise or exercise, but also occurred at rest [73]. The burnout syndrome has been associated with AF in some families [74]. The burnout syndrome type 2 has been associated with two mutations in the gene *kcnq1* (KvLQT1 protein) [30] which leads to a gain of potassium channel function, leading to a shortening of action potential with AF. There is a particular entity, by affecting the same gene that is expressed *in uterus* in the form of bradycardia in the neonatal period was diagnosed as AF and SQTS [75]. The burnout syndrome type 3 has been associated with a mutation in the gene *kcnj2* (Kir2.1 protein) located on chromosome 17, involving a speeding of phase 3 of action potential, resulting in a gain of function [76]. The association of BrS pattern with a shorter QT interval has been associated with mutations in the gene *cacna1c* that cause a change in the unit α calcium channel L-type inducing a loss of channel function related to the association to BrS and QT interval shorter than normal, with autosomal dominant inheritance pattern. With the same phenotype and pattern of inheritance, *cacnb2b* mutations causes a change in the unit-type calcium channels, L, giving the association BrS and shortened QT interval [77]. Recently, it has been published a relation between *cacna2d1* and SQTS [78]. However, more studies must be performed to establish a clear clinical-genetic association.

How to Follow a Child with Short QT Syndrome?

Being a rare entity, there are no clinical guidelines for the monitoring of these patients. Quinidine has been tested as a treatment to try to prolong the QT. In cases associated with AF, has raised the use of antiplatelet drugs to prevent embolic complications. The medication does not reverse or control AF.

In cases where the ventricular response is too slow, we recommend the implantation of a pacemaker. Given the high risk of malignant arrhythmias, may consider the possible inducibility of arrhythmias and defibrillator implantation, although the young age of these patients have created much controversy plating in the management. As SQTS is a highly malignant disease, the genetic study of family members is necessary, and baseline ECG should be performed in all patients [79].

Brugada Syndrome

Brugada syndrome (BrS) [80], is characterized by an ECG pattern consisting of coved-type ST-segment elevation in atypical right-bundle branch block in leads V1 to V3 (often referred to as type-1 Brugada ECG pattern) without structural heart disease. BrS is also characterized by an increased risk for SCD resulting from episodes of polymorphic ventricular tachyarrhythmias with an incidence of about 26-38/100.000 person-year. Although the average age of onset of events is about 40 years, SD can affect people of all ages, especially men (75%). Of the patients, between 20% -50% had family history of SD [81].

The penetrance and expressivity of the disorder are highly variable; BrS is generally considered a disorder involving young male adults, with an arrhythmogenic symptom first occurring at age of 40 years, and SCD typically occurring during sleep, particularly in Asia, where it manifests as sudden unexpected nocturnal death syndrome (SUNDS), the most common cause of natural death in young Asians [82]. The ECG pattern can be present at baseline or it may be intermittent. In the latter it and can be unmasked during a drug challenge with a sodium channel blocker (ajmaline or flecainide in children). The description of acute inducers of the ECG pattern is of paramount importance, as some individuals may be at risk during anesthesia, when they take some oral medications (antidepressants or antiarrhythmics) or, of special importance in children, during a febrile episode. Patients with BrS are at risk for life threatening tachyarrhythmias. After surviving a cardiac arrest or the occurrence of syncope, the only treatment having any proven effect on the prevention of SD is the implantable cardioverter-defibrillator (ICD) [81].

To date, 12 genes have been associated to BrS (Table1). Approximately 20% to 30% of patients with BrS have a mutation in the *scn5a* gene, classified as BrS type 1. The *scn5a* gene (alpha subunit of the cardiac sodium channel) is responsible for the phase 0 of the cardiac action potential, a key player in the cardiac electrical activity. Mutations in *scn5a* result in "loss of function" of the sodium channel [83]. Another sodium channel that has been reported to induce BrS is *gpd1-l* gene. It has been shown that mutation of the *gpd1-l* gene reduces the surface membrane expression and reduces the inward sodium current. In addition, *gpd1-l* has been shown to be the cause of some of the SCD in nursing infants [84, 85]. Additionally, it has been published mutations in *scn1b* (sodium channel beta-1 subunit) and *scn3b* (sodium channel beta-3 subunit) also associated to BrS [86, 87]. In the heart, β1-subunit modifies Nav1.5 increasing $I_{Na.}$ The mutation described in *scn3b* alters Nav1.5 trafficking, decreasing $I_{Na.}$ Other gen associated to BrS is *kcne3* which codify MiRP2 protein (β-subunit that regulates the potassium channel I_{to}). The KCNE peptides modulate some potassium currents in the heart. The *kcne3* gene encodes the regulatory subunit of the potassium channel Ito. The relationship between mutations in this gene and the BrS were detected in a Danish family

[88]. Mutation in the *cacna1c* gene is responsible for a defective a unit of the type-L calcium channels. This induces a loss of channel function, linked to the combination of BrS with shorter QT interval. Transmission follows an autosomal dominant pattern. With the same phenotype, mutation of the *cacnb2b* gene leads to a defect in the L-type calcium channel, giving rise to a combination of BrS and shorter QT interval [77]. In 2009 was associated BrS to *hcn4* gene [89], that codifies for HCN4 channel or If channel. The HCN4 channel controls the heart rate, and its mutations also predispose to inherited sick sinus syndrome and LQTS associated with bradycardia. Another gene described associated to BrS is *kcnj8,* also previously related to early repolarization syndrome (ERS) [90]. The report implicate *kcnj8* as a novel J-wave syndrome susceptibility gene and a marker gain of function in the cardiac $K_{(ATP)}$ Kir6.1 channel [91]. Recently, Kattygnarath et al, [92] published a study supporting that *mog1* gene can impair the trafficking of Nav1.5 to the membrane, leading to INa reduction and clinical manifestation of BrS. Also in 2011, Giudicessi et al. provide the first molecular and functional evidence implicating novel *kcnd3* gain-of-function mutations (Kir4.3 protein) in the pathogenesis and phenotypic expression of BrS, with the potential for a lethal arrhythmia being precipitated by a genetically enhanced I_{to} current gradient within the right ventricle where *kcnd3* gene expression is the highest [92]. Finally, and also in 2011, it is published a study recommending *kcne5* gene screening for BrS or IVF affected patients. Therefore, *kcne5* gene modulates I_{to} and its novel variants appeared to cause IVF especially BrS in male patients through gain-of-function effects on I_{to} [93].

How to Follow a Patient with Brugada Syndrome?

Type 1 Brugada ECG pattern is rare in children. Genetic testing will help to diagnose these kids. When genetic testing is not available, children being at risk of BrS (brother/sister or son/daughter of a parent affected with BrS), should be followed as if they had the syndrome, thus, avoidance of Brugada inducing drugs is mandatory (www.brugadadrugs.com).

As fever is one of the promoters of the Brugada pattern, fever control is highly recommended.

We perform flecainide test around the age of 8 years. If it is positive, we perform an electrophysiological study to assess the inducibility of arrhythmias and possible ICD implantation. We are aware of the concerns of implanting a defibrillator in a child.

In patients with BrS and presence of syncope, aborted SD, nocturnal agonal breathing or seizures, is indicated for ICD implantation. If they also show arrhythmias, treatment is indicated with Quinidine.

Competitive exercise is contraindicated competition, but low-intensity sports are allowed [94-96].

Particularly challenging is treatment of electrical storm in BrS patients, which has been anecdotally treated with isoproterenol, disopyramide [97], orciprenaline [98] and quinine [99].

Special should be considered that BrS can associate AV conduction disturbances and supraventricular tachycardia, so we question the presence of palpitations and eventually treat these arrhythmias with ablation.

Study of the Relatives of a Patient with Brugada Syndrome

In patients with BrS, as in other channelopathies, good family history is crucial. All first grade relatives should be screened by a basal ECG. When in doubt, flecainide test was performed.

In the case of a known mutation, family members should be studied to rule out that they are carriers despite the negativity of the basic ECG and flecainide test.

Lev-Lenègre Syndrome

Lev-Lenègre syndrome is a rare entity characterized by disruption of the conduction system, in which a block gradually develops, resulting in ventricular arrhythmias or asystolia [100, 101]. The amount of sodium and the speed with which it enters the cell determine the velocity of conduction of the electric impulse through the sodium-dependent cells (muscle cells of the ventricle and atrium and cells of the His-Purkinje system). If a mutation leads to a reduction in the quantity of sodium that enters the cell, the velocity of conduction of the impulse is reduced resulting in a loss of function in phase 0 of the action potential (channel opening). This is the case in the Lev-Lenègre syndrome. In 1995, chromosomal abnormalities (19q13.2-13.3) associated with bundle branch block were reported [102]. In 1999 the first mutation was described, located on the *scn5a* gene [103, 104]. Additionally, it has been described mutations in *nkx2.5* gene (5q35) that codify the transcription factor NKX2.5 (also named CSX) [105]. The conduction block is due to a congenital heart defect. Recently, it has also been described a mutation in *trpm4* gene (19q) [106]. The *trpm4* gene is a causative gene in isolated cardiac conduction disease with mutations resulting in a gain of function and TRPM4 channel being highly expressed in cardiac Purkinje fibers (Table1).

Catecholaminergic Polymorphic Ventricular Tachycardia

In 1975 was first reported Catecholaminergic Polymorphic Ventricular Tachycardia (CPVT) [107]. CPVT is a familial arrhythmogenic disorder characterized by a 2-way polymorphic ventricular tachycardia characterized by severe arrhythmias in young patients with apparently normal hearts [108]. It is triggered exclusively by adrenergic stimulus (vigorous exercise, fear), and has a high mortality (30% by the age of 30 years) [109, 110]. When doubt exists, it is recommended to perform and exercise ECG and Holter monitoring in order to rule out the bidirectional tachycardia. The CPVT is associated with a normal ECG at rest (occasionally with bradycardia, and U waves). It was thought that the event happened in childhood (before age 10) but recent studies show that the first manifestation may occur from infancy to age 40 [111].

Three genetic variants have been identified (Table1), an autosomal dominant one caused by mutation in the gene encoding the ryanodine receptor *ryr2* and a recessive one, caused by mutation in the calciquestrina isoform gene *(casq2)* [112, 113]. Both genes are implicated in regulating intracellular calcium and both types of defect lead to increased function of these proteins, and so outflow of calcium from the sarcoplasmatic reticulum is increased. This excess calcium is associated with abnormalities in the sarcolemmal membrane potential, leading to late depolarizations that cause a predisposition to arrhythmias [17]. Similarly, some patients diagnosed with CPVT type 3 on the basis of the presence of bidirectional ventricular tachycardia on exercise have been identified as possessing *kcnj2* mutations, which are associated with the rarely lethal Andersen-Tawil syndrome (ATS1, LQT7) [114]. The misdiagnosis of Andersen-Tawil syndrome as the potentially lethal disorder CPVT may lead to a more aggressive prophylactic therapy (ie, implantation of an ICD) than necessary. Genetic testing may provide a clear differential diagnosis between atypical LQT1 and CPVT and between CPVT and ATS1 [115].

Untreated, the mortality rate is very high, reaching 30-50% in young adulthood. The earlier episodes appear, the worse the prognosis and there is a correlation between the age at which syncope occurs for the first time and the severity of the disease. The first line treatment in patients with CPVT is beta-blockers, which have significantly reduced syncope and SCD [79]. Given that the first symptom may be the SCD, it is recommended to treat all individuals genetically identified as mutation carriers and patients who are asymptomatic but who have ventricular arrhythmias during exercise. The sport is contraindicated, including patients treated with beta-blockers [116].

How to Follow a Child with Catecholaminergic Polymorphic Ventricular Tachycardia?

CPVT should be suspected in any form of dizziness, syncope or aborted SD in a child not only during exercise but also after an episode of fear or running after a bus. We may see sinus bradycardia at the resting ECG, with or without U waves.

When CPVT is suspected, exercise test up to maximum effort may give the diagnostic, but it should be carefully performed. Any type of ventricular premature beat or couples should be carefully observed, as the following phase may show non-sustained ventricular tachycardia, and sustained ventricular tachycardia, that may give the diagnostic. Physical activity is completely contraindicated in these patients [117].

Beta-blockers at very high doses should be given. The dose is adjusted according to the medical tolerance to these drugs. If arrhythmias persist, verapamil can be added verapamil. Cardiac sympathectomy has been suggested to control CPVT in patients refractory to beta-blockers [118], although relief after this surgical procedure can be temporary or of delayed onset [119].

Flecainide suppressed delayed after-depolarization (DADs) in mutant Purkinje fibers [120] and effectively prevented CPVT in mice and humans [121-124]. Defibrillator is indicated in case of aborted SD and in those cases of uncontrolled ventricular tachycardias. Holter is held every three months in order to assess ventricular premature beat.

Study of the Relatives of a Patient with CPVT

In patients with catecholaminergic polymorphic ventricular tachycardia is essential to study all family members basal ECG and exercise testing. It is advisable that the genetic study of these patients. In the case of a known mutation, family members should be studied to rule out that they are carriers.

Idiopathic Ventricular Fibrillation

IVF – spontaneous ventricular fibrillation without identifiable structural or electrical heart disease – may account for up to 10% of SD in the young [125-127]. Haplotype-sharing analysis has identified a genetic basis for IVF [128]. The shared chromosomal segment contained the *dpp6* gene, which encodes a putative regulator of the transient outward Ito current [129]. DPP6 mRNA levels were increased 20-fold in hearts of human carriers in one study [129]. To date, this seems to be a founder risk locus, but nonetheless it suggests that an increase in DPP6 imparts a higher risk for ventricular fibrillation (VF).

Furthermore, previously thought to be a benign and common ECG finding present in up to 5% of the population, three separate case–control studies [130-132] suggest that 'J-point elevation' (manifested either as terminal QRS slurring or notching, or ST-segment elevation with upper concavity and prominent T waves in inferolateral leads) is significantly more prevalent (16–60%) in patients with IVF. Mutations in genes encoding subunits of the L-type Ca2 channel (*cacna1c, cacnb2,* and *cacna2d1*) [133] and a subunit of the KATP channel encoded by *kcnj8* [90] have been implicated in this new 'J-wave syndrome' or ERS [134].

CARDIOMYOPATHIES

Cardiomyopathies are heart diseases induced by mutations in genes that encode contractile and structural proteins as well as proteins for cardiac energy production. They are responsible for lethal arrhythmogenic disorders, mainly hypertrophic cardiomyopathy (HCM) and arrhythmogenic cardiomyopathy (arrhythmogenic right ventricular dysplasia-ARVD/C-), both associated to SCD [135].

Hypertrophic Cardiomyopathy

HCM is one of the most common cardiovascular disorders, self characterized by an unexplained asymmetric hypertrophy of the left ventricle with findings indicative of myocyte disarray and fibrosis [136], often mostly pronounced in the interventricular septum. It has a prevalence of 1/500, affecting children and young people [137], and it is also cause of death in about a 10% of the SCD identified at autopsy [12].

Clinical manifestations appear initially as diastolic dysfunction and systolic-diastolic dysfunction in more advanced stages. Thus, a patient may be asymptomatic or present with SCD. Mortality is higher in young patients (often athletes) than in adults, and the first manifestation of the disease may be SCD itself. The risk stratification of SCD in patients with HCM remains at the forefront of clinical research. Both, a wall thickness of the left ventricle with a z-score>6 and an abnormal blood pressure response to exercise are clinical factors for SCD in children. The differential diagnosis between HCM and so-called "athlete's heart" is not easy, but testing such as ECG, echocardiogram and genetic analysis are useful tools to resolve the dilemma [95, 96, 138, 139]. Patients considered at high risk of SCD should be aimed at prevention and/or management of ventricular tachyarrhythmias with antiarrhythmic drugs (amiodarone). ICD implantation is the most reasonable option for patients at greatest risk. Special emphasis should be put on risk stratification in childhood because evidence points to a higher risk of SCD in this population than in adults [140, 141]. Competitive exercise is an absolute contraindication for children with HCM, leisure activities, however, are possible if the cardiac function permits.

The disease is considered inherited in 90% of the cases, generally with an autosomal dominant pattern of transmission (Table 2), except for cases with mutations in mitochondrial DNA (mtDNA), which are inherited from the mother [142, 143]. Mutations have been described in several genes encoding essential sarcomeric proteins, such as heavy chain β-myosin (*myh7*) and myosin binding protein C (*mybpc3*); others encode heavy chain a-myosin (*myh6*), troponin I (*tnni3*), troponin T (*tnnt2*), α-tropomyosin (*tpm1*), essential myosin light chains (*myl3*), regulatory light chain (*myl2*), titin (*ttn*) and α-actin (*actc*) [144]. Mutations have also been detected in genes implicated in the metabolism of the heme and Fe2+ group, and in genes involved in mitochondrial bioenergetics. Genetic studies of families with left ventricular hypertrophy have shown metabolic myocardiopathies with mutations in the *prkag2* and *lamp2* genes. Recently, missense mutations that cause defective interaction between nexilin and α-actin have been described in HCM [145]. Nexilin (*nexn*) is a cardiac Z-disc protein that has a crucial function to protect cardiac Z-discs from forces generated within the sarcomere [145].

Table 2. Hypertrophic Cardiomyopathy

DISEASE	LOCUS	GENE
Hypertrophic Cardiomyopathy (HCM)	14q11.2-q12	*MYH6*
	14q11.2-q12	*MYH7*
	11p11.2	*MYBPC3*
	12q23-q24.3	*MYL2*
	3p21.2-p21.3	*MYL3*
	1q32	*TNNT2*
	19q13.4	*TNNI3*
	15q22.1	*TPM1*
	15q14	*ACTC*
	2q24.3	*TTN*
	3p21-p24	*TNNC1*
	11p15.1	*CRP3*
	17q12	*TCAP*
	19q13.11	*CALR3*
	6q22.1	*PHOSPHOLAMBAN (PLN)*
	1q42.1-q43	*RYR2*
	20q12	*JUNCTOPHILIN-2 (JPH2)*
	10q22.1-q23	*METAVINCULIN (VCL)*
	17q12-q21.1	*TELETHONIN (TCAP)*
	4q26-q27	*MYOZENIN-2 (MYOZ2)*
	10q22.2-q23.3	*LBD3*
	11p15.1	*CSRP3*
	1q42-q43	*α-ACTININ-2 (ACTN2)*
	9q13	*FXN*
	Xq22	*GLA*
	Xq24	*LAMP2*
	7q35-q36.36	*PRKAG2*
	3p25.2	*RAF1*
	1p31.1	*NEXN*

Up until present, mutations have not been thought to predict phenotype because individuals with different degrees of hypertrophy or with a greater predisposition to SD may be present in the same family and have the same mutation. This is due to the intervention of modifying genes and polymorphisms, which require more exhaustive studies to achieve a full understanding. It is assumed that interruption of mitochondrial energy metabolism in the heart is the cause of HCM in patients with problems of sarcomeric contraction; this sheds some light on several clinical observations such as heterogeneity, variability in clinical presentation, and asymmetry in hypertrophy. The identification of the genotype may contribute to risk stratification, but future genotype-phenotype studies need to be done to confirm whether this is useful.

How to Follow a Child with Hypertrophic Cardiomyopathy

The sport is clearly contraindicated and any isometric exercise (lifting weights). In all patients we recommend the administration of beta-blockers in high doses. The dose is adjusted according to the medical tolerance to these drugs [146-149].

The implantation of an ICD will depend on the presence of symptoms, septal thickness, presence of family history of arrhythmias and SD.

Study of the Relatives of a Patient with Hypertrophic Cardiomyopathy

In patients with HCM is important to perform echocardiography and ECG to their relatives. It is advisable that the genetic study of these patients. In the case of a known mutation, family members should be studied to rule out that they are carriers.

Arrhythmogenic Right Ventricular Dysplasia/Cardiomyopathy

Arrhythmogenic right ventricular dysplasia/cardiomyopathy (ARVD/C) is an inherited cardiomyopathy characterized by right ventricular dysfunction and ventricular arrhythmias. Patients with ARVD/C show fibrofatty replacement of right ventricular infarction. However, ARVD/C may involve both ventricles and in very isolated cases only the left ventricle. The ECG is not always altered, but the presence of epsilon waves and T negative ε, from V1 to V3, may provide a pathological diagnosis [150]. About 50% of ARVD/C cases are familiar with variable penetrance. It affects approximately 1/5000 individuals although the prevalence is higher in men (80%), particularly young athletes. Most cases are diagnosed before the age of 40 years. Usually, the individuals affected have symptomatic ventricular arrhythmias that originate in the right ventricle, with syncope and a high risk of SD. This entity is responsible for 5% of all SCD.

The disease has 2 different patterns of transmission: autosomal dominant pattern (the most common) and an autosomal recessive pattern (Table 3). The recessive pattern has been reported on the Greek island (Naxos), giving rise to the Naxos syndrome. This syndrome comprises ARVD/C, palmoplantar keratoderma and typically curly hair. To date, several genes are described as capable of inducing disease (http://www.arvcdatabase.info/). Most cases involve genes encoding proteins responsible for the junctions of intercalated discs, especially in desmosome, which led to the appointment of ARVD/C as a desmosomal disease with a disorder of the connections between myocytes [151].

The clinical picture may include (a) a subclinical stage hidden structural defects, during which the affected person may have a cardiac arrest/SCD as the first manifestation of the disease, (b) an electrical disorder with palpitations and syncope tachyarrhythmias arising the right ventricle, often triggered during stress, and (c) the failure of the pump of the right ventricle, sometimes severe enough to require a heart transplant [152].

Table 3. Arrhythmogenic Right Ventricular Dysplasia/Cardiomyopathy

DISEASE	LOCUS	GENE	PROTEIN
Arrhythmogenic Right Ventricular Dysplasia Cardiomyopathy (ARVD/C)	14q23-q24	*TGFB3*	Transforming growth factor beta 3
	14q12-q22	-	-
	2q32.1-q32.3	-	-
	3q21.3-3p23	*TMEM43*	Transmembrain protein 43
	10p12p14	-	-
	10q22.3	-	-
	6p24	*DSP*	Desmoplakin
	12p11	*PKP2*	Plakofilin 2
	18q12.1-q12.2	*DSG2*	Desmoglein 2
	18q21	*DSC2*	Desmocollin 2
	17q21	*JUP*	Plakoglobin
	2q35	*DES*	Desmin
	3q27	*TP63*	Tumor protein p63

How to Follow a Child with ARVD/C?

Imaging tests such as echocardiography (presence of fat, impaired mobility, hypokinesia, aneurysms and hypertrabeculated) and MRI (late gadolinium enhancement) allow us to see signs of disease when either patent. Ultrasound and normal ECG does not rule out ARVD/C.

The Holter will guide us as a criterion for dysplasia when there are >500 premature ventricular contractions. The exercise test demonstrated the presence of extrasystoles.

Endomyocardial biopsy is not indicated as it is a patchy disease and appears epicardium to endocardium, so that a negative result we do not rule out the presence of the disease.

These patients should be treated with the highest dose sotalol to allow us to maintain the QT below 360ms. The sport is clearly discouraged. In patients with positive genotype without structural alteration recommendation is restrict physical activity but withhold sotalol.

Study of the Relatives of a Patient with Dysplasia

In patients with ARVD/C is important to have a complete family history, and do a baseline ECG to make immediate family and a stress test. It is advisable that the genetic study of these patients. In the case of a known mutation, family members should be studied to rule out that they are carriers.

Dilated Cardiomyopathy

Dilated cardiomyopathy (DCM) is characterized by left ventricular dilatation and left ventricular systolic dysfunction. Right ventricular dilatation and dysfunction may also be present but are not necessary for the diagnosis. Patients present signs of heart failure or palpitations [153, 154] and SCD are not a frequent manifestation [59]. There are many factors that can trigger DCM, making it a highly heterogeneous entity. Despite this, systematic studies show that at least 35% of cases are hereditary. Arrhythmias that occur in patients with familial DCM are usually the same as in the acquired forms, with defects in atrioventricular and intraventricular conduction, ventricular arrhythmias and AF. Normally, a progressive decline in ventricular function occurs in these patients and they die of either heart failure or arrhythmic events.

In adults, the prevalence is 1/2500 individuals, with an incidence of 7/100000 per year (but it could be underdiagnosed). This disorder develops at any age, in either sex, and in people of any ethnic origin. In adults, DCM arises more commonly in men than in women. In children, the yearly incidence is 0,57 cases per 100000 per year overall, but is higher in boys than in girls (0·66 vs. 0·47 cases per 100000, $p<0·006$), in black people than in white people (0,98 vs. 0,46 cases per 100000, $p<0·001$), and in babies younger than 1 year than in children (4,40 vs. 0,34 cases per 100000, $p<0·001$). Two-thirds of children are thought to have idiopathic disease.

DCM is an extremely complex disease, and so the clinical usefulness of genetic analysis is limited. Even so, systematic studies of relatives of patients with DCM indicate that at least 35% of the cases are hereditary (Table 5). In 1994, the first locus of DCM with atrioventricular block was identified on chromosome 1. To date, there are mutations in genes that code for proteins of the cytoskeleton, cell nucleus and sarcomere [155, 156]. In 30% of the cases, the mutation occurs in the *lamin A/C* gene (*lmna*), which encodes a protein that is expressed in almost all cell types and whose function is to contribute to the integrity of the nucleus by providing mechanical support. Other genes, such as *myh7, mybpc3, tnni3, tnnc1, tcap, vcl, csrp3, pln, ttn, tpm1, actc* and *tnnt2* can also cause DCM. Moreover, a mutation in the *scn5a* gene was identified in a large family with DCM [157]. A recent new addition to the long list of DCM candidate genes is nebulette (*nebl*), which encodes a 107 kDa protein that aligns thin filaments and connects them with the myocardial Z-disk [158].

Another is dynamin-1-like (*dnm1l*) gene, which is known to be critical for mitochondrial fission and has also been implicated in causing DCM by reducing levels of mitochondrial enzyme complexes and cardiac ATP depletion [159].

It is estimated that 20%-50% of the DCM cases are heredity. Three patterns of transmission are described in the inherited forms:

a) Autosomal dominant disease, a locus for which has been reported in several genes (*actin, desmin, lamin A/C, δ-sarcoglycan* and *β-sarcoglycan*);
b) X-linked disease, associated to mutations in *dystrophin* and *tafazin*. Although mutations in the *dystrophin* gene are not a common cause of DCM, direct mutations in this gene give rise to Duchene or Becker-type muscular dystrophy, which affect both cardiac and skeletal muscle;
c) Mitochondrial diseases, which typically affect other tissues besides the myocardium; so far, 2 loci of DCM have been reported in association with primary arrhythmias, one not being the cause of the other. These families had autosomal dominant disease.

DCM is associated with complex remodeling of one or both ventricles, resulting in a change of the ventricle shape and the architecture of the myocardium fibers. In the most severe cases, affected individuals present signs and symptoms of heart failure—diaphoresis, breathlessness at rest or with exertion, orthopnoea, exercise intolerance, early onset fatigue, abdominal pain, and pallor. Heart failure symptoms can be exercise-induced or persistent at rest.

Table 4. LV noncompaction cardiomyopathy

DISEASE	LOCUS	GENE	INHERITANCE
LV noncompaction cardiomyopathy (LVNC)	14q11.2-q13	*MYH7*	Autosomic dominant
	1q32	*TNNT2*	Autosomic dominant
	15q22.1	*TPM1*	Autosomic dominant
	Xq28	*TAFAZZIN (TAZ)*	X-linked
	1q21	*LAMIN A/C (LMNA)*	Autosomic dominant
	10q22.2-q23.3	*LDB3*	Autosomic dominant
	18q12	*DTNA*	Autosomic dominant

Conduct a Patient with Dilated Cardiomyopathy

The degree of dilatation limits physical activity. Patients should be treated by heart failure, regardless of the presence or absence of arrhythmias. If there dyssynchrony, assess the implementation of a resynchronization. In those cases in which malignant arrhythmias are induced, implantation of an ICD is recommended [59, 160, 161].

Study of the Relatives of a Patient with Dilated Cardiomyopathy

In patients with DCM, echocardiography should be performed to their families. It is important to have a good history of family history and make a family tree. It is advisable that the genetic study of these patients [162]. In the case of a known mutation, family members should be studied to rule out that they are carriers [163].

Left Ventricular Noncompaction

Left ventricular noncompaction (LVNC) – a rare disorder characterized by multiple deep trabeculations within the left ventricular myocardium – is increasingly being recognized as a cause of ventricular tachyarrhythmia and SCD in young patients [164]. Overlapping genetic predisposition to LVNC, HCM, and DCM, suggesting a continuum of disease associated with sarcomeric gene mutations, has been proposed [165].

Interestingly, distal chromosome deletion at 22q11.2 [166] and 1p36 [167, 168], though previously associated with atrial and ventricular septal defects, has been more recently described in patients with LVNC (Table 4).

Table 5. Dilated cardiomyopathy

DISEASE	LOCUS	GENE	INHERITANCE
Dilated Cardiomyopathy (DCM)	14q11.2-q13	*MYH7*	Autosomic dominant
	14q12	*MYH6*	Autosomic dominant
	3p21	*MYL3*	Autosomic dominant
	12q23	*MYL2*	Autosomic dominant
	11p11.2	*MYBPC3*	Autosomic dominant
	9q13-q22	*CMD1B*	Autosomic dominant
	9q22-q31	*SEMA4D*	Autosomic dominant
	10q22.1	*MYPN*	Autosomic dominant
	10q22.3-23.2	*ZASP/CYPHER (LDB3)*	Autosomic dominant
	15q11-q14	*ACTC1*	Autosomic dominant
	1q31.3	*TNNI1*	Autosomic dominant
	1q32	*TNNT2*	Autosomic dominant
	19q13	*TNNI3*	Autosomic recessive
	3p21.1	*TNNC1*	Autosomic dominant
	1q42-q43	*α2-ACTININ (ACTN2)*	Autosomic dominant
	15q22.1	*TPM1*	Autosomic dominant
	2q31	*TITIN (TTN)*	Autosomic dominant
	11p15.1	*CSRP3*	Autosomic dominant
	17q12-q21.1	*TELETHONIN (TCAP)*	Autosomic dominant
	Xp21.2	*DYSTROPHIN (DMD)*	X-linked
	2q35	*DESMIN (DES)*	Autosomic dominant
	10q22.1-q23	*METAVINCULIN (VCL)*	Autosomic dominant
	10q23.22	*ANKRD1*	Autosomic dominant
	10q25.3	*RBM20*	Autosomic dominant
	4q12	*β-SARCOGLYCAN (SGCB)*	Autosomic dominant
	17q12	*A-SACOGLYCAN (SGCA)*	Autosomic dominant
	5q33	*δ-SARCOGLYCAN (SGCD)*	Autosomic dominant
	13q12	*γ-SARCOGLYCAN (SGCG)*	Autosomic dominant
	Xq28	*TAFAZZIN (TAZ)(G4.5)*	X-linked
	1q21	*LAMIN A/C (LMNA)*	Autosomic dominant
	6q12-q16	*CMD1K*	Autosomic dominant
	6q22.1	*PHOSPHOLAMBAN (PLN)*	Autosomic dominant
	3p21	*SCN5A*	Autosomic dominant
	5q31	*TTID/MYOT*	Autosomic dominant
	6q13	*MYO6*	Autosomic dominant
	19q13.3	*FKRP*	Autosomic dominant
	11q22.3-23.1	*CRYAB*	Autosomic dominant
	Xq28	*EMD*	X-linked
	6q25	*SYNE1*	Autosomic dominant
	6q23	*EYA4*	Autosomic recessive
	12q22	*TMPO*	Autosomic dominant
	12p12.1	*ABCC9*	Autosomic dominant
	12p11	*PKP2*	Autosomic dominant
	6q24	*DSP*	Autosomic recessive
	18q12.1	*DSG2*	Autosomic dominant
	17q21	*JUP*	Autosomic dominant
	18q12.1	*TTR*	Autosomic dominant
	2p23	*LCHAD/HADHA*	Autosomic dominant
	6p21.3	*HFE*	Autosomic dominant
	1p21	*AGL*	Autosomic dominant
	Xq22	*GLA*	Autosomic dominant
	3p12	*GBE*	X-linked
	Xq24	*LAMP2*	Autosomic dominant
	10q24	*COX15*	X-linked
	12q22	*LAP2/TMPO*	Autosomic dominant
	10p12	*NEBL*	Autosomic dominant
	12p11.21	*DNM1L*	Autosomic dominant

OTHER ANOMALIES WITH GENETIC DEFECTS

Anomalous Origin of the Coronary Arteries

Several congenital coronary artery malformations are the second cause of SD in young athletes, accounting for 12% to 20% of SCD cases [169-172]. The anomalous origin of the left main coronary artery from the right (anterior) sinus of Valsalva, with the left coronary artery coursing between the aorta and the pulmonary trunk to reach the left atrioventricular groove, is the most common type found at autopsies of subjects with SD or death due to anatomical coronary artery anomalies [173]. During exertion, increased stroke volume causes expansion of both the aorta and the pulmonary trunk, which can lead to further compression of the already slit like ostial lumen of the anomalous artery [174]. Decreased blood flow and resultant decreased O_2 supply, in combination with increased O_2 demand during exercise, can lead to myocardial ischemia. This ischemia can trigger ventricular arrhythmias and subsequent SD [175].

Most patients with coronary artery anomalies are asymptomatic, and those who are symptomatic rarely complain of typical angina, reporting nonspecific symptoms such as syncope, dyspnea on exertion and palpitations [174]. Transthoracic echocardiography is utilized in diagnosis although highly trained ultrasound technicians using high-quality imaging systems are required for accurate and consistent identification of the origin of the coronary arteries [174]. Genetic predisposition for these defects has been identified, but diagnostic is made by echocardiographic findings and if exist, it should be corrected surgically.

Marfan Syndrome

Marfan syndrome (MS) has an increased risk for SD due to aortic dissection [176]. Aortic root dilation and dissection are secondary to pathological changes in the aorta, namely, cystic medial necrosis [177]. The disease is an autosomal dominant genetic disease with variable penetrance, with an estimated incidence of 1 in 7000. It induce particular characteristics as tall thin children, with arachnid fingers, long face, associated with strong myopic defects, cardiac valve prolepses and other connective tissue anomalies [176, 178]. More than 100 mutations on the *fibrillin-1* gene have been identified as causes of MS. These mutations lead to defective fibrillin in the extracellular matrix, resulting in abnormalities of the ocular, cardiovascular, skeletal, pulmonary, and integumentary systems [176]. The Berlin nosology or the more recent "Ghent criteria", which consider family history of MS, skeletal and cardiovascular features, as well as other phenotypic expressions, are used to aid in diagnosis [179]. These characteristics may be subtle, and body habitus measurements may be useful. Echocardiography can be used to measure and monitor the degree of aortic root dilation. Non-Marfan cases of familial aortic root aneurysms are also now being recognized and studied [177].

Loeys-Dietz Syndrome

Loeys-Dietz syndrome (LDS) is a rare autosomal dominant disorder showing the involvement of cutaneous, cardiovascular, craniofacial, and skeletal systems. In particular, LDS patients show arterial tortuosity with widespread vascular aneurysm and dissection, and have a high risk of aortic dissection or rupture at an early age and at aortic diameters that ordinarily are not predictive of these events [180, 181]. The LDS is a severe form of extracellular matrix disease, with more premature SD. It has an autosomal dominant inheritance pattern, related to a mutation in either the *tgfbr1* or *tgfbr2* genes (transforming growth factor beta receptor 1 or 2). The LDS has been subdivided in type I (LDSI) and type II (LDSII) on the basis of the presence or the absence of cranio-facial involvement, respectively. Furthermore, LDSII patients display at least two of the major signs of vascular Ehlers-Danlos syndrome (EDS) [182]. These genes encode the receptors for a molecule that plays a role in extracellular signaling. A wide variety of mutations have been identified in both genes, however, there is no significant correlation between the mutation, the location, the gene, and clinical presentation [183-185]. There is no way to predict the severity of vascular, skeletal or skin findings that may occur in an offspring.

REFERENCES

[1] Krous HF, Beckwith JB, Byard RW, et al. Sudden infant death syndrome and unclassified sudden infant deaths: a definitional and diagnostic approach. *Pediatrics.* Jul 2004;114(1):234-238.

[2] Byard RW, Krous HF. Research and sudden infant death syndrome: definitions, diagnostic difficulties and discrepancies. *J. Paediatr. Child Health.* Aug 2004;40(8):419-421.

[3] Papadakis M, Sharma S, Cox S, et al. The magnitude of sudden cardiac death in the young: a death certificate-based review in England and Wales. *Europace.* Oct 2009;11(10):1353-1358.

[4] Wisten A, Forsberg H, Krantz P, et al. Sudden cardiac death in 15-35-year olds in Sweden during 1992-99. *J. Intern. Med.* Dec 2002;252(6):529-536.

[5] Vaartjes I, Hendrix A, Hertogh EM, et al. Sudden death in persons younger than 40 years of age: incidence and causes. *Eur. J. Cardiovasc. Prev. Rehabil.* Oct 2009;16(5):592-596.

[6] Cote A, Russo P, Michaud J. Sudden unexpected deaths in infancy: what are the causes? *J. Pediatr.* Oct 1999;135(4):437-443.

[7] Wren C. Sudden death in children and adolescents. *Heart.* Oct 2002;88(4):426-431.

[8] Arnestad M, Vege A, Rognum TO. Evaluation of diagnostic tools applied in the examination of sudden unexpected deaths in infancy and early childhood. *Forensic. Sci. Int.* Feb 18 2002;125(2-3):262-268.

[9] Liberthson RR. Sudden death from cardiac causes in children and young adults. *N. Engl. J. Med.* Apr 18 1996;334(16):1039-1044.

[10] Guntheroth WG, Spiers PS. The triple risk hypotheses in sudden infant death syndrome. *Pediatrics.* Nov 2002;110(5):e64.

[11] Liberthson RR, Dinsmore RE, Fallon JT. Aberrant coronary artery origin from the aorta. Report of 18 patients, review of literature and delineation of natural history and management. *Circulation.* Apr 1979;59(4):748-754.

[12] Ferrero-Miliani L, Holst AG, Pehrson S, et al. Strategy for clinical evaluation and screening of sudden cardiac death relatives. *Fundam. Clin. Pharmacol.* Oct 2010;24(5):619-635.

[13] Mohler PJ, Wehrens XH. Mechanisms of human arrhythmia syndromes: abnormal cardiac macromolecular interactions. *Physiology (Bethesda).* Oct 2007;22:342-350.

[14] Priori SG, Napolitano C. Role of genetic analyses in cardiology: part I: mendelian diseases: cardiac channelopathies. *Circulation.* Feb 28 2006;113(8):1130-1135.

[15] Schwartz PJ. Cardiac sympathetic innervation and the sudden infant death syndrome. A possible pathogenetic link. *Am. J. Med.* Feb 1976;60(2):167-172.

[16] Maron BJ, Clark CE, Goldstein RE, et al. Potential role of QT interval prolongation in sudden infant death syndrome. *Circulation.* Sep 1976;54(3):423-430.

[17] Klaver EC, Versluijs GM, Wilders R. Cardiac ion channel mutations in the sudden infant death syndrome. *Int. J. Cardiol.* Jan 5 2011.

[18] Tester DJ, Ackerman MJ. Cardiomyopathic and channelopathic causes of sudden unexplained death in infants and children. *Annu. Rev. Med.* 2009;60:69-84.

[19] Moss AJ, Kass RS. Long QT syndrome: from channels to cardiac arrhythmias. *J. Clin. Invest.* Aug 2005;115(8):2018-2024.

[20] Goldenberg I, Moss AJ. Long QT syndrome. *J. Am. Coll Cardiol.* Jun 17 2008;51(24):2291-2300.

[21] Viskin S, Halkin A. Treating the long-QT syndrome in the era of implantable defibrillators. *Circulation.* Jan 20 2009;119(2):204-206.

[22] Schimpf R, Veltmann C, Wolpert C, et al. Channelopathies: Brugada syndrome, long QT syndrome, short QT syndrome, and CPVT. *Herz.* Jun 2009;34(4):281-288.

[23] Schimpf R, Veltmann C, Wolpert C, et al. Arrhythmogenic hereditary syndromes: Brugada Syndrome, long QT syndrome, short QT syndrome and CPVT. *Minerva Cardioangiol.* Dec 2010;58(6):623-636.

[24] Arnestad M, Crotti L, Rognum TO, et al. Prevalence of long-QT syndrome gene variants in sudden infant death syndrome. *Circulation.* Jan 23 2007;115(3):361-367.

[25] Otagiri T, Kijima K, Osawa M, et al. Cardiac ion channel gene mutations in sudden infant death syndrome. *Pediatr. Res.* Nov 2008;64(5):482-487.

[26] Roden DM, Viswanathan PC. Genetics of acquired long QT syndrome. *J. Clin. Invest.* Aug 2005;115(8):2025-2032.

[27] Tester DJ, Benton AJ, Train L, et al. Prevalence and spectrum of large deletions or duplications in the major long QT syndrome-susceptibility genes and implications for long QT syndrome genetic testing. *Am. J. Cardiol.* Oct 15 2010;106(8):1124-1128.

[28] Kapplinger JD, Tester DJ, Salisbury BA, et al. Spectrum and prevalence of mutations from the first 2,500 consecutive unrelated patients referred for the FAMILION long QT syndrome genetic test. *Heart Rhythm.* Sep 2009;6(9):1297-1303.

[29] Chen S, Zhang L, Bryant RM, et al. KCNQ1 mutations in patients with a family history of lethal cardiac arrhythmias and sudden death. *Clin. Genet.* Apr 2003;63(4):273-282.

[30] Bellocq C, van Ginneken AC, Bezzina CR, et al. Mutation in the KCNQ1 gene leading to the short QT-interval syndrome. *Circulation.* May 25 2004;109(20):2394-2397.

[31] Curran ME, Splawski I, Timothy KW, et al. A molecular basis for cardiac arrhythmia: HERG mutations cause long QT syndrome. *Cell.* Mar 10 1995;80(5):795-803.

[32] January CT, Gong Q, Zhou Z. Long QT syndrome: cellular basis and arrhythmia mechanism in LQT2. *J. Cardiovasc. Electrophysiol.* Dec 2000;11(12):1413-1418.

[33] Bianchi L, Shen Z, Dennis AT, et al. Cellular dysfunction of LQT5-minK mutants: abnormalities of IKs, IKr and trafficking in long QT syndrome. *Hum. Mol. Genet.* Aug 1999;8(8):1499-1507.

[34] Tsuboi M, Antzelevitch C. Cellular basis for electrocardiographic and arrhythmic manifestations of Andersen-Tawil syndrome (LQT7). *Heart Rhythm.* Mar 2006;3(3):328-335.

[35] Yang Y, Yang Y, Liang B, et al. Identification of a Kir3.4 Mutation in Congenital Long QT Syndrome. *Am. J. Hum. Genet.* Jun 2 2010.

[36] Schwartz PJ, Priori SG, Locati EH, et al. Long QT syndrome patients with mutations of the SCN5A and HERG genes have differential responses to Na+ channel blockade and to increases in heart rate. Implications for gene-specific therapy. *Circulation.* Dec 15 1995;92(12):3381-3386.

[37] Wang Q, Shen J, Splawski I, et al. SCN5A mutations associated with an inherited cardiac arrhythmia, long QT syndrome. *Cell.* Mar 10 1995;80(5):805-811.

[38] Medeiros-Domingo A, Kaku T, Tester DJ, et al. SCN4B-encoded sodium channel beta4 subunit in congenital long-QT syndrome. *Circulation.* Jul 10 2007;116(2):134-142.

[39] Ueda K, Valdivia C, Medeiros-Domingo A, et al. Syntrophin mutation associated with long QT syndrome through activation of the nNOS-SCN5A macromolecular complex. *Proc. Natl. Acad. Sci. USA.* Jul 8 2008;105(27):9355-9360.

[40] Wu G, Ai T, Kim JJ, et al. alpha-1-syntrophin mutation and the long-QT syndrome: a disease of sodium channel disruption. *Circ. Arrhythm Electrophysiol.* Aug 2008;1(3):193-201.

[41] Vatta M, Ackerman MJ, Ye B, et al. Mutant caveolin-3 induces persistent late sodium current and is associated with long-QT syndrome. *Circulation.* Nov 14 2006;114(20):2104-2112.

[42] Splawski I, Timothy KW, Sharpe LM, et al. Ca(V)1.2 calcium channel dysfunction causes a multisystem disorder including arrhythmia and autism. *Cell.* Oct 1 2004;119(1):19-31.

[43] Yarotskyy V, Gao G, Peterson BZ, et al. The Timothy syndrome mutation of cardiac CaV1.2 (L-type) channels: multiple altered gating mechanisms and pharmacological restoration of inactivation. *J. Physiol.* Feb 1 2009;587(Pt 3):551-565.

[44] Kauferstein S, Kiehne N, Erkapic D, et al. A novel mutation in the cardiac ryanodine receptor gene (RyR2) in a patient with an unequivocal LQTS. *Int. J. Cardiol.* Jan 21 2011;146(2):249-250.

[45] Chen L, Marquardt ML, Tester DJ, et al. Mutation of an A-kinase-anchoring protein causes long-QT syndrome. *Proc. Natl. Acad. Sci. USA.* Dec 26 2007;104(52):20990-20995.

[46] Mohler PJ. Ankyrins and human disease: what the electrophysiologist should know. *J. Cardiovasc. Electrophysiol.* Oct 2006;17(10):1153-1159.

[47] Mohler PJ, Bennett V. Ankyrin-based cardiac arrhythmias: a new class of channelopathies due to loss of cellular targeting. *Curr. Opin. Cardiol.* May 2005;20(3):189-193.

[48] Mohler PJ, Schott JJ, Gramolini AO, et al. Ankyrin-B mutation causes type 4 long-QT cardiac arrhythmia and sudden cardiac death. *Nature.* Feb 6 2003;421(6923):634-639.
[49] Moss AJ, Zareba W, Hall WJ, et al. Effectiveness and limitations of beta-blocker therapy in congenital long-QT syndrome. *Circulation.* Feb 15 2000;101(6):616-623.
[50] Goldenberg I, Moss AJ, Peterson DR, et al. Risk factors for aborted cardiac arrest and sudden cardiac death in children with the congenital long-QT syndrome. *Circulation.* Apr 29 2008;117(17):2184-2191.
[51] Hobbs JB, Peterson DR, Moss AJ, et al. Risk of aborted cardiac arrest or sudden cardiac death during adolescence in the long-QT syndrome. *Jama.* Sep 13 2006;296(10):1249-1254.
[52] Chatrath R, Bell CM, Ackerman MJ. Beta-blocker therapy failures in symptomatic probands with genotyped long-QT syndrome. *Pediatr. Cardiol.* Sep-Oct 2004;25(5):459-465.
[53] Schwartz PJ, Priori SG, Cerrone M, et al. Left cardiac sympathetic denervation in the management of high-risk patients affected by the long-QT syndrome. *Circulation.* Apr 20 2004;109(15):1826-1833.
[54] Pellizzon OA, Kalaizich L, Ptacek LJ, et al. Flecainide suppresses bidirectional ventricular tachycardia and reverses tachycardia-induced cardiomyopathy in Andersen-Tawil syndrome. *J. Cardiovasc. Electrophysiol.* Jan 2008;19(1):95-97.
[55] Schulze-Bahr E, Fenge H, Etzrodt D, et al. Long QT syndrome and life threatening arrhythmia in a newborn: molecular diagnosis and treatment response. *Heart.* Jan 2004;90(1):13-16.
[56] Moltedo JM, Kim JJ, Friedman RA, et al. Use of a cardioselective beta-blocker for pediatric patients with prolonged QT syndrome. *Pediatr. Cardiol.* Jan 2011;32(1):63-66.
[57] Roden DM. Clinical practice. Long-QT syndrome. *N. Engl. J. Med.* Jan 10 2008;358(2):169-176.
[58] Schwartz PJ, Spazzolini C, Priori SG, et al. Who are the long-QT syndrome patients who receive an implantable cardioverter-defibrillator and what happens to them?: data from the European Long-QT Syndrome Implantable Cardioverter-Defibrillator (LQTS ICD) Registry. *Circulation.* Sep 28 2010;122(13):1272-1282.
[59] Zipes DP, Camm AJ, Borggrefe M, et al. ACC/AHA/ESC 2006 guidelines for management of patients with ventricular arrhythmias and the prevention of sudden cardiac death--executive summary: A report of the American College of Cardiology/American Heart Association Task Force and the European Society of Cardiology Committee for Practice Guidelines (Writing Committee to Develop Guidelines for Management of Patients with Ventricular Arrhythmias and the Prevention of Sudden Cardiac Death) Developed in collaboration with the European Heart Rhythm Association and the Heart Rhythm Society. *Eur. Heart J.* Sep 2006;27(17):2099-2140.
[60] Hofbeck M, Ulmer H, Beinder E, et al. Prenatal findings in patients with prolonged QT interval in the neonatal period. *Heart.* Mar 1997;77(3):198-204.
[61] Garson A, Jr., Dick M, 2nd, Fournier A, et al. The long QT syndrome in children. An international study of 287 patients. *Circulation.* Jun 1993;87(6):1866-1872.

[62] Gorgels AP, Al Fadley F, Zaman L, et al. The long QT syndrome with impaired atrioventricular conduction: a malignant variant in infants. *J. Cardiovasc. Electrophysiol.* Nov 1998;9(11):1225-1232.
[63] Horigome H, Nagashima M, Sumitomo N, et al. Clinical characteristics and genetic background of congenital long-QT syndrome diagnosed in fetal, neonatal, and infantile life: a nationwide questionnaire survey in Japan. *Circ. Arrhythm. Electrophysiol.* Feb 1 2010;3(1):10-17.
[64] Barsheshet A, Peterson DR, Moss AJ, et al. Genotype-Specific QT Correction for Heart Rate and the Risk of Life Threatening Cardiac Events in Adolescents with the Congenital Long-QT Syndrome. *Heart Rhythm.* Mar 9 2011.
[65] Gussak I, Brugada P, Brugada J, et al. Idiopathic short QT interval: a new clinical syndrome? *Cardiology.* 2000;94(2):99-102.
[66] Morita H, Wu J, Zipes DP. The QT syndromes: long and short. *Lancet.* Aug 30 2008;372(9640):750-763.
[67] Crotti L, Taravelli E, Girardengo G, et al. Congenital short QT syndrome. *Indian Pacing Electrophysiol J.* 2010;10(2):86-95.
[68] Bjerregaard P, Nallapaneni H, Gussak I. Short QT interval in clinical practice. *J. Electrocardiol.* Sep-Oct 2010;43(5):390-395.
[69] Triedman JK. Brugada and short QT syndromes. *Pacing Clin. Electrophysiol.* Jul 2009;32 Suppl 2:S58-62.
[70] Zareba W, Cygankiewicz I. Long QT syndrome and short QT syndrome. *Prog Cardiovasc Dis.* Nov-Dec 2008;51(3):264-278.
[71] Patel C, Yan GX, Antzelevitch C. Short QT syndrome: from bench to bedside. *Circ. Arrhythm. Electrophysiol.* Aug 1 2010;3(4):401-408.
[72] Patel U, Pavri BB. Short QT syndrome: a review. *Cardiol. Rev.* Nov-Dec 2009;17(6):300-303.
[73] Brugada R, Hong K, Dumaine R, et al. Sudden death associated with short-QT syndrome linked to mutations in HERG. *Circulation.* Jan 6 2004;109(1):30-35.
[74] Hong K, Bjerregaard P, Gussak I, et al. Short QT syndrome and atrial fibrillation caused by mutation in KCNH2. *J. Cardiovasc. Electrophysiol.* Apr 2005;16(4):394-396.
[75] Hong K, Piper DR, Diaz-Valdecantos A, et al. De novo KCNQ1 mutation responsible for atrial fibrillation and short QT syndrome in utero. *Cardiovasc. Res.* Dec 1 2005;68(3):433-440.
[76] Priori SG, Pandit SV, Rivolta I, et al. A novel form of short QT syndrome (SQT3) is caused by a mutation in the KCNJ2 gene. *Circ. Res.* Apr 15 2005;96(7):800-807.
[77] Antzelevitch C, Pollevick GD, Cordeiro JM, et al. Loss-of-function mutations in the cardiac calcium channel underlie a new clinical entity characterized by ST-segment elevation, short QT intervals, and sudden cardiac death. *Circulation.* Jan 30 2007;115(4):442-449.
[78] Templin C, Ghadri JR, Rougier JS, et al. Identification of a novel loss-of-function calcium channel gene mutation in short QT syndrome (SQTS6). *Eur. Heart J.* Mar 10 2011.
[79] Zipes DP, Camm AJ, Borggrefe M, et al. ACC/AHA/ESC 2006 guidelines for management of patients with ventricular arrhythmias and the prevention of sudden cardiac death: a report of the American College of Cardiology/American Heart

Association Task Force and the European Society of Cardiology Committee for Practice Guidelines (Writing Committee to Develop Guidelines for Management of Patients With Ventricular Arrhythmias and the Prevention of Sudden Cardiac Death). *J. Am. Coll Cardiol.* Sep 5 2006;48(5):e247-346.

[80] Brugada P, Brugada J. Right bundle branch block, persistent ST segment elevation and sudden cardiac death: a distinct clinical and electrocardiographic syndrome. A multicenter report. *J. Am. Coll Cardiol.* Nov 15 1992;20(6):1391-1396.

[81] Benito B, Brugada J, Brugada R, et al. Brugada syndrome. *Rev. Esp. Cardiol.* Nov 2009;62(11):1297-1315.

[82] Vatta M, Dumaine R, Varghese G, et al. Genetic and biophysical basis of sudden unexplained nocturnal death syndrome (SUNDS), a disease allelic to Brugada syndrome. *Hum. Mol. Genet.* Feb 1 2002;11(3):337-345.

[83] Kapplinger JD, Tester DJ, Alders M, et al. An international compendium of mutations in the SCN5A-encoded cardiac sodium channel in patients referred for Brugada syndrome genetic testing. *Heart Rhythm.* Jan 2010;7(1):33-46.

[84] London B, Michalec M, Mehdi H, et al. Mutation in glycerol-3-phosphate dehydrogenase 1 like gene (GPD1-L) decreases cardiac Na+ current and causes inherited arrhythmias. *Circulation.* Nov 13 2007;116(20):2260-2268.

[85] Van Norstrand DW, Valdivia CR, Tester DJ, et al. Molecular and functional characterization of novel glycerol-3-phosphate dehydrogenase 1 like gene (GPD1-L) mutations in sudden infant death syndrome. *Circulation.* Nov 13 2007;116(20):2253-2259.

[86] Watanabe H, Koopmann TT, Le Scouarnec S, et al. Sodium channel beta1 subunit mutations associated with Brugada syndrome and cardiac conduction disease in humans. *J. Clin. Invest.* Jun 2008;118(6):2260-2268.

[87] Hu D, Barajas-Martinez H, Burashnikov E, et al. A mutation in the beta 3 subunit of the cardiac sodium channel associated with Brugada ECG phenotype. *Circ. Cardiovasc. Genet.* Jun 2009;2(3):270-278.

[88] Delpon E, Cordeiro JM, Nunez L, et al. Functional effects of KCNE3 mutation and its role in the development of Brugada syndrome. *Circ. Arrhythm. Electrophysiol.* Aug 2008;1(3):209-218.

[89] Ueda K, Hirano Y, Higashiuesato Y, et al. Role of HCN4 channel in preventing ventricular arrhythmia. *J. Hum. Genet.* Feb 2009;54(2):115-121.

[90] Haissaguerre M, Chatel S, Sacher F, et al. Ventricular fibrillation with prominent early repolarization associated with a rare variant of KCNJ8/KATP channel. *J. Cardiovasc. Electrophysiol.* Jan 2009;20(1):93-98.

[91] Medeiros-Domingo A, Tan BH, Crotti L, et al. Gain-of-function mutation S422L in the KCNJ8-encoded cardiac K(ATP) channel Kir6.1 as a pathogenic substrate for J-wave syndromes. *Heart Rhythm.* Oct 2010;7(10):1466-1471.

[92] Kattygnarath D, Maugenre S, Neyroud N, et al. MOG1: A New Susceptibility Gene for Brugada Syndrome. *Circ. Cardiovasc. Genet.* Mar 29 2011.

[93] Ohno S, Zankov DP, Ding WG, et al. KCNE5 (KCNE1L) Variants Are Novel Modulators of Brugada Syndrome and Idiopathic Ventricular Fibrillation. *Circ. Arrhythm. Electrophysiol.* Jun 1 2011;4(3):352-361.

[94] Corrado D, Basso C, Pavei A, et al. Trends in sudden cardiovascular death in young competitive athletes after implementation of a preparticipation screening program. *Jama.* Oct 4 2006;296(13):1593-1601.
[95] Maron BJ. Distinguishing hypertrophic cardiomyopathy from athlete's heart physiological remodelling: clinical significance, diagnostic strategies and implications for preparticipation screening. *Br. J. Sports Med.* Sep 2009;43(9):649-656.
[96] Maron BJ, Doerer JJ, Haas TS, et al. Sudden deaths in young competitive athletes: analysis of 1866 deaths in the United States, 1980-2006. *Circulation.* Mar 3 2009;119(8):1085-1092.
[97] Sumi S, Maruyama S, Shiga Y, et al. High efficacy of disopyramide in the management of ventricular fibrillation storms in a patient with Brugada syndrome. *Pacing Clin. Electrophysiol.* Jun 1 2010;33(6):e53-56.
[98] Schweizer PA, Becker R, Katus HA, et al. Successful acute and long-term management of electrical storm in Brugada syndrome using orciprenaline and quinine/quinidine. *Clin. Res. Cardiol.* Jul 2010;99(7):467-470.
[99] Mehrotra S, Juneja R, Naik N, et al. Successful Use of Quinine in the Treatment of Electrical Storm in a Child with Brugada Syndrome. *J. Cardiovasc. Electrophysiol.* Oct 6 2010.
[100] Lenegre J, Moreau P. [Chronic Auriculo-Ventricular Block. Anatomical, Clinical and Histological Study.]. *Arch. Mal. Coeur. Vaiss.* Aug 1963;56:867-888.
[101] Lev M. Anatomic Basis for Atrioventricular Block. *Am. J. Med.* Nov 1964;37:742-748.
[102] Brink PA, Ferreira A, Moolman JC, et al. Gene for progressive familial heart block type I maps to chromosome 19q13. *Circulation.* Mar 15 1995;91(6):1633-1640.
[103] Tan HL, Bink-Boelkens MT, Bezzina CR, et al. A sodium-channel mutation causes isolated cardiac conduction disease. *Nature.* Feb 22 2001;409(6823):1043-1047.
[104] Schott JJ, Alshinawi C, Kyndt F, et al. Cardiac conduction defects associate with mutations in SCN5A. *Nat. Genet.* Sep 1999;23(1):20-21.
[105] Schott JJ, Benson DW, Basson CT, et al. Congenital heart disease caused by mutations in the transcription factor NKX2-5. *Science.* Jul 3 1998;281(5373):108-111.
[106] Liu H, El Zein L, Kruse M, et al. Gain-of-Function Mutations in TRPM4 Cause Autosomal Dominant Isolated Cardiac Conduction Disease. *Circ. Cardiovasc. Genet.* Jun 19 2010.
[107] Reid DS, Tynan M, Braidwood L, et al. Bidirectional tachycardia in a child. A study using His bundle electrography. *Br. Heart J.* Mar 1975;37(3):339-344.
[108] Viskin S, Belhassen B. Polymorphic ventricular tachyarrhythmias in the absence of organic heart disease: classification, differential diagnosis, and implications for therapy. *Prog. Cardiovasc. Dis.* Jul-Aug 1998;41(1):17-34.
[109] Hayashi M, Denjoy I, Extramiana F, et al. Incidence and risk factors of arrhythmic events in catecholaminergic polymorphic ventricular tachycardia. *Circulation.* May 12 2009;119(18):2426-2434.
[110] Ylanen K, Poutanen T, Hiippala A, et al. Catecholaminergic polymorphic ventricular tachycardia. *Eur. J. Pediatr.* May 2010;169(5):535-542.
[111] Laitinen PJ, Swan H, Piippo K, et al. Genes, exercise and sudden death: molecular basis of familial catecholaminergic polymorphic ventricular tachycardia. *Ann. Med.* 2004;36 Suppl 1:81-86.

[112] Medeiros-Domingo A, Iturralde-Torres P, Ackerman MJ. [Clinical and genetic characteristics of long QT syndrome]. *Rev. Esp. Cardiol.* Jul 2007;60(7):739-752.
[113] Lahat H, Eldar M, Levy-Nissenbaum E, et al. Autosomal recessive catecholamine- or exercise-induced polymorphic ventricular tachycardia: clinical features and assignment of the disease gene to chromosome 1p13-21. *Circulation.* Jun 12 2001;103(23):2822-2827.
[114] Tester DJ, Arya P, Will M, et al. Genotypic heterogeneity and phenotypic mimicry among unrelated patients referred for catecholaminergic polymorphic ventricular tachycardia genetic testing. *Heart Rhythm.* Jul 2006;3(7):800-805.
[115] Tester DJ, Ackerman MJ. Genetic testing for potentially lethal, highly treatable inherited cardiomyopathies/channelopathies in clinical practice. *Circulation.* Mar 8 2011;123(9):1021-1037.
[116] Katz G, Arad M, Eldar M. Catecholaminergic polymorphic ventricular tachycardia from bedside to bench and beyond. *Curr. Probl. Cardiol.* Jan 2009;34(1):9-43.
[117] Leenhardt A, Lucet V, Denjoy I, et al. Catecholaminergic polymorphic ventricular tachycardia in children. A 7-year follow-up of 21 patients. *Circulation.* Mar 1 1995;91(5):1512-1519.
[118] Wilde AA, Bhuiyan ZA, Crotti L, et al. Left cardiac sympathetic denervation for catecholaminergic polymorphic ventricular tachycardia. *N. Engl. J. Med.* May 8 2008;358(19):2024-2029.
[119] Gopinathannair R, Olshansky B, Iannettoni M, et al. Delayed maximal response to left cardiac sympathectomy for catecholaminergic polymorphic ventricular tachycardia. *Europace.* Jul 2010;12(7):1035-1039.
[120] Kang G, Giovannone SF, Liu N, et al. Purkinje cells from RyR2 mutant mice are highly arrhythmogenic but responsive to targeted therapy. *Circ. Res.* Aug 20 2010;107(4):512-519.
[121] Watanabe H, Chopra N, Laver D, et al. Flecainide prevents catecholaminergic polymorphic ventricular tachycardia in mice and humans. *Nat. Med.* Apr 2009;15(4):380-383.
[122] Hilliard FA, Steele DS, Laver D, et al. Flecainide inhibits arrhythmogenic Ca2+ waves by open state block of ryanodine receptor Ca2+ release channels and reduction of Ca2+ spark mass. *J. Mol. Cell Cardiol.* Feb 2010;48(2):293-301.
[123] Knollmann BC. Power and pitfalls of using transgenic mice to optimize therapy for CPVT: a need for prospective placebo-controlled clinical trials in genetic arrhythmia disorders. *Heart Rhythm.* Nov 2010;7(11):1683-1685.
[124] Biernacka EK, Hoffman P. Efficacy of flecainide in a patient with catecholaminergic polymorphic ventricular tachycardia. *Europace.* Jan 2011;13(1):129-130.
[125] van der Werf C, van Langen IM, Wilde AA. Sudden death in the young: what do we know about it and how to prevent? *Circ. Arrhythm. Electrophysiol.* Feb 1 2010;3(1):96-104.
[126] Survivors of out-of-hospital cardiac arrest with apparently normal heart. Need for definition and standardized clinical evaluation. Consensus Statement of the Joint Steering Committees of the Unexplained Cardiac Arrest Registry of Europe and of the Idiopathic Ventricular Fibrillation Registry of the United States. *Circulation.* Jan 7 1997;95(1):265-272.

[127] van der Werf C, Hofman N, Tan HL, et al. Diagnostic yield in sudden unexplained death and aborted cardiac arrest in the young: the experience of a tertiary referral center in The Netherlands. *Heart Rhythm.* Oct 2010;7(10):1383-1389.

[128] Alders M, Koopmann TT, Christiaans I, et al. Haplotype-sharing analysis implicates chromosome 7q36 harboring DPP6 in familial idiopathic ventricular fibrillation. *Am. J. Hum. Genet.* Apr 2009; 84(4):468-476.

[129] Radicke S, Cotella D, Graf EM, et al. Expression and function of dipeptidyl-aminopeptidase-like protein 6 as a putative beta-subunit of human cardiac transient outward current encoded by Kv4.3. *J. Physiol.* Jun 15 2005;565(Pt 3):751-756.

[130] Haissaguerre M, Derval N, Sacher F, et al. Sudden cardiac arrest associated with early repolarization. *N. Engl. J. Med.* May 8 2008; 358(19):2016-2023.

[131] Nam GB, Kim YH, Antzelevitch C. Augmentation of J waves and electrical storms in patients with early repolarization. *N. Engl. J. Med.* May 8 2008;358(19):2078-2079.

[132] Rosso R, Kogan E, Belhassen B, et al. J-point elevation in survivors of primary ventricular fibrillation and matched control subjects: incidence and clinical significance. *J. Am. Coll Cardiol.* Oct 7 2008;52(15):1231-1238.

[133] Burashnikov E, Pfeiffer R, Barajas-Martinez H, et al. Mutations in the cardiac L-type calcium channel associated with inherited J-wave syndromes and sudden cardiac death. *Heart Rhythm.* Dec 2010;7(12):1872-1882.

[134] Miyazaki S, Shah AJ, Haissaguerre M. Early repolarization syndrome - a new electrical disorder associated with sudden cardiac death. *Circ. J.* Oct 2010;74(10):2039-2044.

[135] Mitten MJ, Maron BJ, Zipes DP. Task Force 12: legal aspects of the 36th Bethesda Conference recommendations. *J. Am. Coll Cardiol.* Apr 19 2005;45(8):1373-1375.

[136] Basso C, Calabrese F, Corrado D, et al. Postmortem diagnosis in sudden cardiac death victims: macroscopic, microscopic and molecular findings. *Cardiovasc. Res.* May 2001;50(2):290-300.

[137] Ho CY. Hypertrophic cardiomyopathy. *Heart Fail Clin.* Apr 2010;6(2):141-159.

[138] Pelliccia A. Athlete"s heart and hypertrophic cardiomyopathy. *Curr. Cardiol. Rep.* Mar 2000;2(2):166-171.

[139] Sharma S, Maron BJ, Whyte G, et al. Physiologic limits of left ventricular hypertrophy in elite junior athletes: relevance to differential diagnosis of athlete's heart and hypertrophic cardiomyopathy. *J. Am. Coll Cardiol.* Oct 16 2002;40(8):1431-1436.

[140] Ostman-Smith I, Wettrell G, Keeton B, et al. Age- and gender-specific mortality rates in childhood hypertrophic cardiomyopathy. *Eur. Heart J.* May 2008;29(9):1160-1167.

[141] Ostman-Smith I, Wettrell G, Keeton B, et al. Echocardiographic and electrocardiographic identification of those children with hypertrophic cardiomyopathy who should be considered at high-risk of dying suddenly. *Cardiol. Young.* Dec 2005;15(6):632-642.

[142] Wang L, Seidman JG, Seidman CE. Narrative review: harnessing molecular genetics for the diagnosis and management of hypertrophic cardiomyopathy. *Ann. Intern. Med.* Apr 20 2010;152(8):513-520, W181.

[143] Marian AJ. Hypertrophic cardiomyopathy: from genetics to treatment. *Eur. J. Clin. Invest.* Apr 2010;40(4):360-369.

[144] Konno T, Chang S, Seidman JG, et al. Genetics of hypertrophic cardiomyopathy. *Curr. Opin. Cardiol.* Jan 30 2010.

[145] Wang H, Li Z, Wang J, et al. Mutations in NEXN, a Z-disc gene, are associated with hypertrophic cardiomyopathy. *Am. J. Hum. Genet.* Nov 12 2010;87(5):687-693.

[146] Corrado D, Basso C, Schiavon M, et al. Screening for hypertrophic cardiomyopathy in young athletes. *N. Engl. J. Med.* Aug 6 1998;339(6):364-369.

[147] Maron BJ, Shirani J, Poliac LC, et al. Sudden death in young competitive athletes. Clinical, demographic, and pathological profiles. *Jama.* Jul 17 1996;276(3):199-204.

[148] Ackerman MJ. Cardiac causes of sudden unexpected death in children and their relationship to seizures and syncope: genetic testing for cardiac electropathies. *Semin. Pediatr. Neurol.* Mar 2005;12(1):52-58.

[149] Ackerman MJ. Genetic testing for risk stratification in hypertrophic cardiomyopathy and long QT syndrome: fact or fiction? *Curr. Opin. Cardiol.* May 2005;20(3):175-181.

[150] Ellinor PT, MacRae CA, Thierfelder L. Arrhythmogenic right ventricular cardiomyopathy. *Heart Fail Clin.* Apr 2010;6(2):161-177.

[151] Awad MM, Calkins H, Judge DP. Mechanisms of disease: molecular genetics of arrhythmogenic right ventricular dysplasia/cardiomyopathy. *Nat. Clin. Pract. Cardiovasc. Med.* May 2008;5(5):258-267.

[152] Azaouagh A, Churzidse S, Konorza T, et al. Arrhythmogenic right ventricular cardiomyopathy/dysplasia: a review and update. *Clin. Res. Cardiol.* Mar 1 2011.

[153] Hamilton RM, Azevedo ER. Sudden cardiac death in dilated cardiomyopathies. *Pacing Clin. Electrophysiol.* Jul 2009;32 Suppl 2:S32-40.

[154] Luk A, Ahn E, Soor GS, et al. Dilated cardiomyopathy: a review. *J. Clin. Pathol.* Mar 2009;62(3):219-225.

[155] Fatkin D, Otway R, Richmond Z. Genetics of dilated cardiomyopathy. *Heart Fail Clin.* Apr 2010;6(2):129-140.

[156] Dellefave L, McNally EM. The genetics of dilated cardiomyopathy. *Curr. Opin. Cardiol.* Feb 24 2010.

[157] Moller DV, Andersen PS, Hedley P, et al. The role of sarcomere gene mutations in patients with idiopathic dilated cardiomyopathy. *Eur. J. Hum. Genet.* Oct 2009;17(10):1241-1249.

[158] Purevjav E, Varela J, Morgado M, et al. Nebulette mutations are associated with dilated cardiomyopathy and endocardial fibroelastosis. *J. Am. Coll Cardiol.* Oct 26 2010;56(18):1493-1502.

[159] Ashrafian H, Docherty L, Leo V, et al. A mutation in the mitochondrial fission gene Dnm1l leads to cardiomyopathy. *PLoS Genet.* Jun 2010;6(6):e1001000.

[160] Priori SG, Aliot E, Blomstrom-Lundqvist C, et al. Task Force on Sudden Cardiac Death of the European Society of Cardiology. *Eur. Heart J.* Aug 2001;22(16):1374-1450.

[161] Priori SG, Aliot E, Blomstrom-Lundqvist C, et al. Task Force on Sudden Cardiac Death, European Society of Cardiology. *Europace.* Jan 2002;4(1):3-18.

[162] Hershberger RE, Lindenfeld J, Mestroni L, et al. Genetic evaluation of cardiomyopathy--a Heart Failure Society of America practice guideline. *J. Card Fail.* Mar 2009;15(2):83-97.

[163] Meune C, Van Berlo JH, Anselme F, et al. Primary prevention of sudden death in patients with lamin A/C gene mutations. *N. Engl. J. Med.* Jan 12 2006;354(2):209-210.

[164] Captur G, Nihoyannopoulos P. Left ventricular non-compaction: genetic heterogeneity, diagnosis and clinical course. *Int. J. Cardiol.* Apr 15 2010;140(2):145-153.

[165] Baldi M, Sgalambro A, Nistri S, et al. [Clinical and genetic features of left ventricular noncompaction: a continuum in cardiomyopathies]. *G. Ital. Cardiol. (Rome).* May 2010;11(5):377-385.

[166] Madan S, Madan-Khetarpal S, Park SC, et al. Left ventricular non-compaction on MRI in a patient with 22q11.2 distal deletion. *Am. J. Med. Genet A.* May 2010;152A(5):1295-1299.

[167] Cremer K, Ludecke HJ, Ruhr F, et al. Left-ventricular non-compaction (LVNC): a clinical feature more often observed in terminal deletion 1p36 than previously expected. *Eur. J. Med. Genet.* Nov-Dec 2008;51(6):685-688.

[168] Gajecka M, Saitta SC, Gentles AJ, et al. Recurrent interstitial 1p36 deletions: Evidence for germline mosaicism and complex rearrangement breakpoints. *Am. J. Med. Genet. A.* Dec 2010;152A(12):3074-3083.

[169] Fuller CM. Cost effectiveness analysis of screening of high school athletes for risk of sudden cardiac death. *Med. Sci. Sports Exerc.* May 2000;32(5):887-890.

[170] Iskandar EG, Thompson PD. Exercise-related sudden death due to an unusual coronary artery anomaly. *Med. Sci. Sports Exerc.* Feb 2004;36(2):180-182.

[171] Maron BJ. Sudden death in young athletes. *N. Engl. J. Med.* Sep 11 2003;349(11):1064-1075.

[172] Trusty JM, Beinborn DS, Jahangir A. Dysrhythmias and the athlete. *AACN Clin. Issues.* Jul-Sep 2004;15(3):432-448.

[173] Maron BJ. Risk profiles and cardiovascular preparticipation screening of competitive athletes. *Cardiol. Clin.* Aug 1997;15(3):473-483.

[174] Zeppilli P, dello Russo A, Santini C, et al. In vivo detection of coronary artery anomalies in asymptomatic athletes by echocardiographic screening. *Chest.* Jul 1998;114(1):89-93.

[175] Loeys B, De Paepe A. New insights in the pathogenesis of aortic aneurysms. *Verh. K. Acad. Geneeskd. Belg.* 2008;70(2):69-84.

[176] Kinoshita N, Mimura J, Obayashi C, et al. Aortic root dilatation among young competitive athletes: echocardiographic screening of 1929 athletes between 15 and 34 years of age. *Am. Heart J.* Apr 2000;139(4):723-728.

[177] Bader RS, Goldberg L, Sahn DJ. Risk of sudden cardiac death in young athletes: which screening strategies are appropriate? *Pediatr. Clin. North Am.* Oct 2004;51(5):1421-1441.

[178] Munoz L, Norgan G, Rauschhuber M, et al. An exploratory study of cardiac health in college athletes. *Appl. Nurs. Res.* Nov 2009;22(4):228-235.

[179] Nienaber CA, Eagle KA. Aortic dissection: new frontiers in diagnosis and management: Part I: from etiology to diagnostic strategies. *Circulation.* Aug 5 2003;108(5):628-635.

[180] Loeys BL, Schwarze U, Holm T, et al. Aneurysm syndromes caused by mutations in the TGF-beta receptor. *N Engl J Med.* Aug 24 2006;355(8):788-798.

[181] Loeys BL, Chen J, Neptune ER, et al. A syndrome of altered cardiovascular, craniofacial, neurocognitive and skeletal development caused by mutations in TGFBR1 or TGFBR2. *Nat. Genet.* Mar 2005;37(3):275-281.

[182] Drera B, Ritelli M, Zoppi N, et al. Loeys-Dietz syndrome type I and type II: clinical findings and novel mutations in two Italian patients. *Orphanet J. Rare Dis.* 2009;4:24.

[183] Gutman G, Baris HN, Hirsch R, et al. Loeys-Dietz syndrome in pregnancy: a case description and report of a novel mutation. *Fetal Diagn Ther.* 2009;26(1):35-37.

[184] Jamsheer A, Henggeler C, Wierzba J, et al. A new sporadic case of early-onset Loeys-Dietz syndrome due to the recurrent mutation p.R528C in the TGFBR2 gene substantiates interindividual clinical variability. *J. Appl. Genet.* 2009;50(4):405-410.

[185] Singh KK, Rommel K, Mishra A, et al. TGFBR1 and TGFBR2 mutations in patients with features of Marfan syndrome and Loeys-Dietz syndrome. *Hum. Mutat.* Aug 2006;27(8):770-777.

In: Comas and Syncope: Causes, Prevention and Treatment ISBN 978-1-62100-603-9
Editors: Eduardo Silva and Gabriel Cruz.

Chapter 3

CARDIOGENIC SYNCOPE

F. R. Breijo-Márquez and M. Pardo Ríos
Murcia University. School of Medicine.
Department of Clinical and Experimental Cardiology. Murcia. Spain

ABSTRACT

Syncope is a transient, brief and sudden loss of consciousness produced by a lessening of the flow of blood to the brain and therefore oxygen deprivation. It can be presented from different ways. Similarly, can be due to a great variety of causes, although the most frequent causes to produce an episode of syncope on any person are: the stroke (embolism and / or brain hemorrhage) and cardiovascular disorders. In many occasions, both are interlinked.

Patients suffer a complete loss of consciousness that may be associated to involuntary movements of the limbs. This may lead to confusion in diagnosis between epilepsy and syncope, while the treatment is quite different for both pathological entities. Unfortunately this confusion is too frequent, so that many cases are treated as epilepsy when in fact is another origin of syncope. Hence, the differential diagnosis should be essential in such events.

All syncope episode gives as result an inexplicable fall of the person who suffers from it.

The most important tool for all physicians for a first diagnostic impression is the patient medical history along with the eyewitness recollection of the incident, if possible. Physicians should be aware of the possibility that a patient or caregiver may inadvertently misinterpret the symptoms when trying to describe to them.

Today is well known that approximately 10% of patients diagnose with epilepsy are syncope sufferers. More than 25% of them have an origin in cardiac abnormalities (this percentage is 45-50% higher when the patient is 60 years or older, regardless of sex).

Any patient with several episodes of syncope or loss of consciousness should be subjected to a full investigation. This should include at least, a blood chemistry analysis, an electroencephalogram (EEF) and an electrocardiogram (ECG), which are fundamental components for evaluation. A normal ECG is a good prognostic feature in most cases.

From the cardiac viewpoint, strictly, the events that can raise concern are:

- Cardiac ischemia and myocardial infarction.
- Any electrical disorder from the heart::

Bradycardia: <40 palpitations per minute (with exception of athletes or pauses of 3 seconds or more).

Atrioventricular block (2° and 3° degrees, especially).

Complete left bundle branch block.

Rapid supraventricular tachycardia (greater than 150 bpm).

Ventricular tachycardia.

Disorders from electrical cardiac systole

- 1°. PR-interval disorders: Wolff-Parkinson-White's Syndrome. Mahaim's Syndrome. Lown-Ganong-Levine's Syndrome.
- 2°. QT-interval disorders: Short QT- interval. Long QT-interval.
- 3° Recording of some ECG patterns like a shortened of " PQ and QT – intervals" together in a same ECG record (Breijo's pattern).
- 4° Brugada Syndrome.
- 5° Other disorders of electrical cardiac conduction.

An ECG without signs suggestive for cardiac pathology should make us think that the syncope is not severe and should be seen as reassuring.

INTRODUCTION

There are two distinct reasons for the evaluation of patients with syncope: one is to identify the precise cause that allows for an effective treatment that be specific for the production mechanism of syncope, the other is the identification of specific risk to the patient, which often depends from underlying disease rather than the mechanism itself syncope [1, 2, 3]. The patient's backgrounds are fundamental to clinicians in order to avoid confusing the two concepts [4,5].

The main objective of this chapter is to develop a summary document concerned to proceed not just to cardiologists, but also to all physicians who are interested in this field.

The most relevant actions to achieve these objectives are listed below:

- An update from the classification of syncope in a broader framework on transient loss of consciousness.
- New epidemiological data.
- A new focused diagnostic approach on risk stratification for sudden cardiac death and cardiovascular events after from the initial assessment, including recommendations for treating patients with syncope of unknown risk.
- A main emphasis for to achieve a better diagnostic strategy based on long-term monitoring, rather than the conventional approach based in laboratory tests only.

As we said in the "abstract", the syncope is a transient and brief loss of consciousness due to a transient global cerebral hypoperfusion and is characterized by a sudden onset, a short duration and a complete spontaneous recovery. In some forms, the syncope can have a prodromal period in which various symptoms (such as auras, nausea, sweating, weakness and visual disturbances) warn that syncope is imminent.

Any sudden and full cessation in cerebral blood flow of just 6-8 seconds has been demonstrated like a sufficient time to cause a complete loss of consciousness.

Classification and Pathophysiology Of Syncope

The common feature of all pathophysiological mechanisms of syncope is a fall in systemic blood pressure, with an overall reduction in cerebral blood flow. Experience, with Tilt table, has shown that a decrease in systolic blood pressure of 60 mm Hg or below is associated with syncope.

Table 1. Classification of Syncope

Reflection (neuromediated) Vasovagal: - Mediated by emotional distress, fear, pain, instrumentation, blood phobia - Mediated by orthostatic stress Facts: - Coughing, sneezing - gastrointestinal stimulation (swallow, defecation, visceral pain) - micturition (post-micturition) - After exercise - Postprandial - Other (eg., laughing, playing wind instruments, lifting weights) Carotid sinus syncope: Atypical forms (without apparent triggers and / or presentation atypical) Syncope due to orthostatic hypotension Primary autonomic dysfunction: - pure primary autonomic dysfunction, multiple system atrophy, Parkinson's disease with autonomic dysfunction, dementia with Lewy corps Secondary autonomic dysfunction: - Diabetes, amyloidosis, uremia, spinal cord injury Drug-induced orthostatic hypotension: - Alcohol, vasodilators, diuretics, phenothiazines, antidepressants Volume depletion: - bleeding, diarrhea, vomiting, etc.
Cardiac syncope

Systemic arterial pressure is determined by cardiac output and full peripheral vascular resistance and a fall in either of these two factors can cause syncope, but often there is a combination of both mechanisms, although their relative contributions can vary considerably. The low or inappropriate peripheral resistance may be due to an inappropriate reflex activity. Other causes of low or inadequate peripheral resistance are functional and structural failure from autonomic nervous system induced by drugs, primary and/or secondary. In the autonomic nervous system dysfunction, the vasomotor sympathetic pathways are unable to increase total peripheral vascular resistance in response to the upright position. When there is a gravitational stress ,combined with the vasomotor sympathetic dysfunction , venous blood accumulation under the diaphragm, causes a decrease in venous return, in cardiac output and, consequently to a reduction in cerebral blood flow with syncopal episode appearance [6, 7,8].

Table 2. Clinical situations incorrectly diagnosed as syncope

Disorders with partial or complete loss of consciousness but not global cerebral hypoperfusion: - Epilepsy -Metabolic disorders, including hypoglycemia, hypoxia, hyperventilation with hypocapnia - Poisoning - Vertebro-basilar transient ischemic attack Disorders with impaired consciousness: - Cataplexy. - Drop Attacks. - Falls. - Functional (psychogenic pseudo-syncope). - Transient ischemic attack of carotid origin.

The causes of transient low cardiac output are threefold. The first is a reflex bradycardia, known as reflex cardio-inhibitory syncope.

The second is from cardiovascular causes, due to structural or non-structural disease such as arrhythmias, arterial hypertension and pulmonary embolism. The third is an inadequate venous return, due to volume depletion or venous pooling.

The ending is three mechanisms: reflex syncope, syncope secondary to orthostatic hypotension and cardiovascular syncope.

In these cases, the differential diagnosis is usually obvious, but sometimes it can be difficult due to the absence of clinical history, recurrent circumstances or confusion about the definition of syncope. This distinction is important for all clinicians who facing to patients with sudden loss of consciousness (real or perceived) that may be due to causes not related to a decrease in the global cerebral blood flow, such as epileptic attacks or a conversion reaction.

Causes of Cardiogenic Syncope

The most frequent syncope‘s episodes from Cardiogenic cause are:

NO STRUCTURAL HEART DISEASE.

The Arrhythmia (is the leading cause):

Bradycardia:

- Sinus node dysfunction (including the syndrome bradycardia / tachycardia).
- Disease of the atrioventricular conduction system.
- Dysfunction of an implantable device

Tachycardia:

- Supraventricular.
- Ventricular (idiopathic or secondary to structural heart disease).

Any channelopathy type as well as any abnormality from electrical cardiac systole.
Bradycardia and tachyarrhythmia, drug-induced
STRUCTURAL HEART DISEASE.
Cardiac valvular disease.
Myocardial infarction / ischemia.
Hypertrophic cardiomyopathy.
Cardiac masses (atrial myxoma, tumors, etc.).
Pericardial disease / tampon.
Congenital anomalies of the coronary arteries.
Prosthetic valve dysfunction.
Other: Pulmonary embolism. Acute aortic dissection. Pulmonary hypertension.

1. Arrhythmias

Any type of cardiac arrhythmia accompanied by important haemodynamic deterioration is capable of cause syncopal episodes. Arrhythmias are the most common causes for cardiac syncope. It produces a haemodynamic deterioration, which can lead to a critical reduction in cardiac output and in cerebral blood flow. Nevertheless, the syncope often has multiple contributing factors, such as heart rate, type of arrhythmia (supraventricular or ventricular), left ventricular function and the adequacy of vascular compensation. This latter includes the neural baroreceptor reflexes and responses to orthostatic hypotension induced by arrhythmia.

Independently of these contributing factors, when the arrhythmia is the primary cause of syncope should be treated specifically.

In the intrinsic sinus node dysfunction, the sinus node is damaged, due to an abnormal automaticity or sinoatrial conduction abnormalities. In this situation, the syncope is caused by a prolonged pauses caused by sinus arrest or sinoatrial block and the failure of the escape mechanisms. These pauses occur often to the final of episodes of atrial tachyarrhythmia (tachycardia-bradycardia syndrome).

The most serious forms of atrioventricular block (AVB) acquired (Mobitz II block, "progressive blocking" and complete atrioventricular block) are related more closely to syncope. In these cases, heart rate may become dependent on pacemakers or subsidiary pacemaker (often unreliable). Syncope occurs, because the pause before these pacemakers begin to shoot is long [9,10]. Besides, these sites act as subsidiary pacemakers and typically have relatively slow frequencies (25-40 bpm). Bradycardia also prolongs ventricular repolarization and predisposes to polymorphic ventricular tachycardia, especially torsade de pointes. Several drugs can cause bradyarrhythmias and tachyarrhythmias. Many antiarrhythmic drugs may produce bradycardia as a result of its specific effect on sinus node function or conduction auriculovaentricular (AV). Syncope due to torsade de pointes is not uncommon, especially in women and is caused by the drugs that prolong the QT interval. It is particularly common in patients with long QT syndrome. Drugs that prolong the QT interval belonging to different categories, such as antiarrhythmics, vasodilators, psycho-tropics, antibiotics, non-sedating antihistamines, etc. Much has been learned on the hereditary long QT syndrome from data collected in the international registry.

2. Structural Cardiac Disease

Structural heart disease can cause syncope when circulatory demand exceeds the limited capacity of the heart to increase cardiac output. Syncope is very worrying when it is associated with clinical situations in which there is a fixed or dynamic obstruction of left ventricular outflow. The basis for the occurrence of an fainting is inadequate blood flow due to mechanical obstruction. However, in some cases, the syncope is not merely the result of restricted cardiac output, but may be partly a reflection of inappropriate or orthostatic hypotension [9-13]. For example, in the context of valvular aortic stenosis, the syncope is not merely the result of restricted cardiac output, but may be due in part to inadequate reflex vasodilatation and / or arrhythmia primary heart. Therefore, the mechanism of syncope may be multifactorial. Recognize in the heart the cause of the problem is justified by the need to correct the underlying structural disease, when possible.

Table 3. Data to be considered as a cardiogenic syncope

- Presence of structural heart disease confirmed
- Family history of sudden cardiac death of unknown cause or channelopathy.
- During the exertion, or supine
- Sudden onset of syncope, palpitations followed immediately.
- Findings indicate the ECG arrhythmic syncope:
- Block bifascicular (defined as blocking both the right and left arm, combined with left anterior fascicular block or left posterior)
- Other intraventricular conduction abnormalities (QRS duration $\geq$ 0.12 s)
- Second-degree atrioventricular block Mobitz I and Mobitz II.
- Complete atrioventricular block.
- Inadequate asymptomatic sinus bradycardia (<50 bpm), sinoatrial block or sinus pause $\geq$ 3 s in the absence of negative chronotropic medications.
- Non-sustained ventricular tachycardia.
- QRS complexes preexcited
- Short, Long QT interval and Short PR interval: Wolff-Parkinson-White's syndrome; Mahaim's syndrome; Lown-Gagnon-Levine's syndrome; Breijo Pattern.
- Early Ventricular Repolarization.
- A pattern of right bundle branch block with ST elevation in leads V1-V3 (Brugada's syndrome).
- Alterations in electrical cardiac systole.
- Negative T waves in right precordial leads, epsilon-waves and ventricular late potentials consistent with arrhythmogenic right ventricular cardiomyopathy.
- Q-waves consistent with myocardial infarction.

EPIDEMIOLOGY AND PROGNOSIS [1, 3,4]

The prevalence of other causes of syncope is different depending on the clinical context in which it evaluates the patient and the patient's age. In addition, there are other differences that depend on the diagnostic definitions, geographical factors and local care pathways, which make comparison between different studies. However, it is possible to make some general comments:

- The reflex syncope is the most frequent syncope in any context.
- Syncope secondary to cardiovascular disease is the second most common cause.

The number of patients with a cardiovascular cause varies widely between studies; the higher frequencies are seen in emergency departments, especially in older patients and consultations to cardiology.

- The orthostatic hypotension is a rare cause of syncope on general population; it is common in very elderly patients.
- Many episodes are misdiagnosed as syncope. Syncopal episodes in the initial evaluation are more frequents in patients referred from emergency and reflect the multifactorial complexity of these patients. While on young people, the reflex syncope is by far the most common cause of transient loss of consciousness, the elderly patients often has multiple causes and medical history may be less reliable than in the young.

It is quite possible that the actual statistics of episodes of syncope can be much higher than today we know. This may be due to a misinterpretation of them. In the clinical practice, we have seen how many of the episodes classified as "epilepsy", in fact, have an etiology due to an electrical cardiac systole's alteration. Especially the so denominated: Syndromes of Lown-Ganong-Levine, Mahaim and Breijo (short PR interval with short QT interval in the same person).

It is possible, therefore, that incidence and prevalence of "Syncope" is very underestimated and can be most important. Recent studies show a "frequency" constant and higher significantly of syncope in the emergency services in Europe, with an incidence of 1% (0.9% -1.7%) of all of cases attended.

In the Framingham study, the incidence of syncope shows a pronounced increase after 70 years of age, from 5.7 episodes/1000 person-years in Males aged 60-69 years to 11.1 / 1,000 in those 70-79 years.

However, in older adults and in elderly patients (> 60 years) the cumulative incidence of syncope throughout the life becomes increasingly difficult to obtain due to bias in data fainting episodes in the past decades. Regarding the prognosis (ie, risk stratification) associated with syncope, one must consider two important elements:

- The risk of death and events that endanger life.
- The risk of recurrence of syncope.

The structural heart disease and primary electrical disease are major risk factors for sudden cardiac death and overall mortality in patients with syncope. Orthostatic hypotension is associated with a risk of death from the severity of co-morbidities that is 2 times the overall population.

In contrast, young patients in who has been excluded the structural and an electrical heart disease, have an excellent prognosis.

The number of patients with a cardiovascular cause varies widely among studies; the higher frequencies are seen in emergency departments, especially in older patients as well as in oriented queries on cardiology.

Most of the deaths and poor prognosis may be related to the severity of the underlying disease, rather than syncope itself. In population studies, approximately one third of patients have recurrence of syncope in 3-year follow-up.

The number of episodes of syncope during the life is the strongest predictor of recurrence.

As premise, we could say:

> "All syncopal episodes that appear on persons under 15 or over 65 should be considered as cardiogenic syncope until proven otherwise".

DIAGNOSIS

The two decision trees that separate the transient loss of consciousness of other clinical situations are based on whether or not unconsciousness and if present the four characteristics that define the presentation of transient loss of consciousness (temporary, quick start , short duration and spontaneous recovery). The transient loss of consciousness is divided in traumatic or no traumatic.

The commotion tends to cause a loss of consciousness; as the existence of traumatism is usually clear, the risk of confusion in diagnosis is small. The transient loss of consciousness is divided in non-traumatic syncope, epileptic attacks, psychogenic pseudo-syncope and a miscellany of more rare causes. Rare causes include both those that are rare (such as cataplexy) and those whose presentation was similar to other forms of transient loss of consciousness only in unusual circumstances (eg., excessive sleepiness during the day).

The initial evaluation of a patient presenting with transient loss of consciousness consists of a careful history, physical examination, determinations include orthostatic blood pressure and ECG [14-19]. Syncope is a common phenomenon in the overall medical practice.

The first thing we must consider is that syncope is not a disease in itself but a symptom. Therefore, it is the specific treatment of primary disease the main aim of any treatment.

Likewise, PREVENTION of syncopal episode must be based on prevention from pathology causing the syncope.

The recurrent vasovagal syncope is the most common diagnosis in this context.

The diagnosis is based on a careful history and the context in which the incident took place. The physician can diagnose most of these common and typical fainting, and just need reassurance. We recommend an active search for worrisome symptoms of syncope as during exercise; syncope while the patient is lying; no external factors; and family history of sudden cardiac death or slow recovery from syncope. If the diagnosis is uncertain and there is a potential risk of harmful consequences, the patient should be referred to a cardiologist, internist, neurologist or psychologist / psychiatrist, as appropriate, or a specialized unit of syncope if available.

Table 5. Differential diagnosis between epilepsy and syncope

Epilepsy can cause transient loss of consciousness: they are "non-responders", who fall and then have amnesia.
This occurs in the atonic seizures, tonic-clonic, clonic and tonic, only. In the epilepsy with absence total on children and in complex partial epilepsy on adults, consciousness is altered but there is not a full loss; patients remain standing during attacks, unlike what happens in the transient loss of consciousness (or syncope).
A complete flaccidity during unconsciousness is against epilepsy. The only exception is the "atonic attacks", but this is rare and occurs without a trigger, on children with pre-existing neurological disease.
Involuntary movements may occur both in epilepsy and in syncope. In epilepsy, the movements are 1 min. aprox. Whereas in syncope are seconds.
The epileptic seizures are large, usually rhythmic and synchronous, whereas syncope is usually asynchronous, small and non-rhythmic. Nevertheless, sometimes the synchronous seizures can occur in syncope and eyewitnesses can report the type of movements incorrectly.
The involuntary movements in syncope occur only after the onset of unconsciousness and after the fall, not so in epilepsy. Syncope is usually triggered, epilepsy, rarely. There are some triggers the reflex epilepsy, such as lights going on and off, different from those of syncope.
A typical aura is a growing feeling in the abdomen (epigastric aura). This growing sense in syncope rarely occurs.
Sweating and pallor are common in epilepsy and no in syncope.
The bite of the language is much more common in epilepsy and occurs on the sides of the tongue, while in syncope occurs at the tip. Urinary incontinence occurs in both.
Patients may feel confused after the attack for some time in epilepsy while syncope is usually retrieved immediately.
Headache, muscle pain and an increase of creatine kinase and prolactin are much more common after epileptic episode.

Management of Syncope in the Overall Practice

Syncope is a common phenomenon in the overall practice. The recurrent vasovagal syncope is the most common diagnosis in this context.

The diagnosis is based on a careful history and the context in which the incident took place. The physician can diagnose most of these common and typical fainting, and just need reassurance. We recommend an active search for worrisome symptoms of syncope as during exercise; syncope while the patient is lying; no external factors; and family history of sudden cardiac death or slow recovery from syncope. If the diagnosis is uncertain and there is a potential risk of harmful consequences, the patient should be referred to a cardiologist, internist, neurologist or psychologist / psychiatrist, as appropriate, or a specialized unit of syncope if available.

Based on the results of different tests, the additional examinations that may be performed [20,21]:

- Carotid sinus massage in patients over 40 years.
- Echocardiogram when is been previously known cardiac disease or findings consistent with structural heart disease or syncope secondary to cardiovascular causes.
- ECG monitoring immediately when suspected arrhythmic syncope.
- Orthostatic tests (proof of inclination or Till test) when the syncope is related to the position vertical or there is suspicion of reflex mechanism.
- Other less specific tests, including neurological examination or blood tests are only appropriate when there is suspicion of transient loss knowledge in non-syncopal.

Management of Cardiac Arrhythmias as Primary Cause for Syncope

We repeat that syncope is not a disease in itself but a consequence of some pathology. In this section would be a result of any cardiac arrhythmia. The aims of treatment are to prevent the recurrence of symptoms, improve quality of life and prolong survival. The basis of syncope in these situations is multifactorial and is influenced by ventricular rate, left ventricular function and the adequacy of vascular compensation (including the potential impact of reflection neuro mediated) [20,21].

1. Sinus Node Dysfunction: In every cases , the cardiac pacemaker therapy is indicated and has proved very effective in patients with sinus node dysfunction when there is an ECG evidence for spontaneous , bradyarrhythmia, that is the cause of syncope or is a result of a recovery time sinus node abnormal. The permanent stimulation often relieves symptoms, but may have no effect on survival. Even when there is adequate stimulation, syncope recurs in 20% of patients in long-term monitoring.

This is due to the frequent association of a vasodepressor reflex mechanism of sinus node disease. We recommended, pacing modes ventricular fibrillation with minimal stimulation have been recently developed as an alternative to conventional rate-dependent pacemaker dual chamber, in patients who mainly require headset support.

The removal of drugs that may exacerbate underlying susceptibility to bradycardia is an important element in preventing the recurrence of syncope. However, when replacement is not feasible may be necessary cardiac stimulation. Cardiac ablation techniques for atrial tachyarrhythmia control have become increasingly important in selected patients presenting as bradycardia-tachycardia, sick sinus syndrome, but are used only occasionally to prevent syncope.

2. Disease Atrioventricular Conduction System:

The pacing is the treatment of syncope associated with symptomatic atrio-ventricular block (AVB).

The indications and the preferential mode of stimulation in the AVB have been updated recently. The possible deleterious role of apical stimulation permanent on right ventricular also been stressed recently, but alternative sites of stimulation are still under discuss.

Stimulation should be considered in patients requiring biventricular pacing due to AVB, depressed Left ventricular ejection , heart failure and wide QRS.

3. Paroxysmal Tachycardias:

Supraventricular and ventricular tachycardia in patients with AV re-entry or typical atrial flutter associated with syncope, treatment of choice is catheter ablation. In these patients, the role of pharmacotherapy is to bridge to the ablation or used when the ablation has failed. In patients with syncope associated with atrial fibrillation or atypical left atrial flutter, the decision must be made individually. Syncope due to torsade de pointes is not uncommon, and is acquired as a result of treatment with drugs that prolong the QT interval. The treatment is to discontinue the drug immediately suspect.

Should be considered for catheter ablation or medical therapy in patients with syncope due to ventricular tachycardia, in the context of a normal heart or heart disease, with mild cardiac dysfunction.

ICDs are indicated in patients with syncope and cardiac function depressed, and tachycardia or ventricular fibrillation without a cause susceptible of correction. Although the ICD patients are usually not prevent recurrence of syncope, is indicated for reducing the risk of sudden cardiac death. Dysfunction of the implanted device 4 oddly shaped implantable stimulation systems have caused pre-syncope or syncope. However, more often than syncope in these patients has no relation with device.

4. Patients with primary electrical disease: The unexplained syncope is considered an ominous finding in patients with inherited disorders of cardiac ion channels. Should be considered carefully the use of an ICD in the absence of other diagnosis or cannot be excluded ventricular tachyarrhythmia as a cause of syncope. However, the mechanism of syncope can be heterogeneous and in some cases, can cause serious arrhythmias, but in many other cases may have a more benign origin, as reflex syncope. Therefore, in these contexts, syncope does not necessarily involve a high risk of major cardiac events that endanger life and is less sensitive than a history of documented cardiac arrest.

In the long QT syndrome, especially in those with LQTS2 and LQTS3, the number of cardiac events before age 18, very long QT intervals and females has a worse clinical outcome [21].

Patients with Brugada syndrome with a spontaneous type 1 ECG pattern have a worse outcome than those who have a pattern or type 2 induced by drugs.

The usefulness of ICD in patients with syncope is controversial and certainly more questionable than in survivors of cardiac arrest [19,20,21].

Special Topics [22,23]

SYNCOPE IN THE ELDERLY.

The most common causes:

Orthostatic hypotension, reflex syncope, especially the carotid sinus syndrome and cardiac arrhythmias.

Often different forms coexist in a patient, which makes diagnosis difficult.

Hospitalization associated with orthostatic hypotension increase progressively with age: 4.2% in patients 65 to 74 years and 30.5% in patients over 75 years.

Symptomatic patients, 25% had orthostatic hypotension 'related age", in the rest, orthostatic hypotension is mainly due to atrial fibrillation medication and primary or secondary.

Systolic hypertension supine often occurs in elderly patients with orthostatic hypotension and complicates treatment, as most of the drugs used to treat orthostatic hypotension exacerbate supine hypertension and vice versa.

SYNCOPE IN PAEDIATRIC PATIENTS

Two specific disorders occur early on in childhood:

1. Infantile reflexes syncopal attacks (also called an oxic apnea or seizures pale reflections) produced by a brief unpleasant stimulus, are due to cardiac vagal inhibition.
2. The transient loss of consciousness type apneic hypoxic (cyanotic apnea also called) is characterized by a cessation of breathing during expiratory crying, cyanosis and often produces transient loss of consciousness.

Other Disorders in Electrical Cardiac Systole as Cause for Syncopal Episodes

The electrical cardiac systole originates from the beginning of the P wave (atrial depolarization) to the end of the descending branch of the T wave (ventricular repolarization). Are included, therefore, the succession of P-QRS-T and its corresponding intervals and segments: PQ, ST and QT. The mathematical possibilities in the variation on length of electrical systole of the heart may be several. It is well documented and demonstrated that such changes in length can cause that be more vulnerable and unstable all myocardial cells, and can also cause serious cardiac arrhythmias, several syncope episodes and even sudden death for this motive. Even today, many of these disorders are poorly understood and, too many times, its clinical manifestations are categorized as "episodes of epilepsy"; other times (most) are classified within a "common sack" called "channelopathies", when -actually- is the alteration from electrical cardiac systole the true etiology of them. All these disorders can cause syncopal episodes.

The measures and lengths of the different components of electrical cardiac systole, considered for most authors as normal are these:

PR-interval: 0.120- 0.200 seconds.

QRS complex: 0.08-0.120 seconds.

QT-interval (corrected): 0.350-0.450 seconds. (Here, there is much disagreement among different authors). The most used methods for QT interval correction, since it is frequency-dependent, are Bazett, Fridericia.

When the PR-interval is lesser than 0.120 seconds, we call it a short PR-interval. In contrast, when is greater than 0,200 seconds, we call it a first-degree AV block. When the QRS complex is lesser than 0.08 seconds, we call it "narrow QRS" but when is greater than

0,120 seconds, we call it "wide QRS". Likewise, when the corrected QT- interval length is lesser than 0,350 seconds, we call it Short QT- interval and when is greater than 0,450 seconds, we call it a Long QTc- interval.

It is clear that there may be, in the same ECG recording, a combination of them all.

Some of these disorders, we will explain briefly below.

A/ Wolff-Parkinson-White's Syndrome (WPWS).

Wolff-Parkinson-White syndrome (WPWS) is a congenital heart disease (PRKAG2. Genetic map 7q36) characterized by a premature ventricular depolarization caused by an abnormal atrioventricular accessory pathway, between the atria and ventricles, known as Kent's bundle. However, even today, is called into question the real cause of Wolff-Parkinson-White, there are some authors who believe that, PRKAG2 mutations, are caused by a glycogen storage cardiomyopathy associated with WPWS, because the overwhelming majority of accessory pathways occur in individuals without structural heart disease, and probably without this mutation. The pathogenesis of accessory pathway formation in PRKAG2 may be completely different, and some authors believe it is due to an inflammation of myocardial cells that occur in the atrial-ventricular connections.

In fact, do not even know if the accessory pathways are mediated genetically or due to environmental exposures or randomly.

A short PR interval, a delta wave, a wide QRS complex (greater than 120 ms) and, occasionally, alterations in the ventricular repolarization are its main electrocardiographic characteristics on the ECG. Its incidence varies between 0, 1% and 3% in the general population.

It is essential to achieve the right differential diagnosis between

- Wolff-Parkinson-White's syndrome or real ventricular pre-excitation.
- Lown-Ganong-Levine syndrome or accelerated atrioventricular conduction.
- Mahaim's syndrome.
- "Short PR alongside short QT" intervals in the same person. (Breijo's Pattern).

Typical ECG image of the Wolff-Parkinson-White.

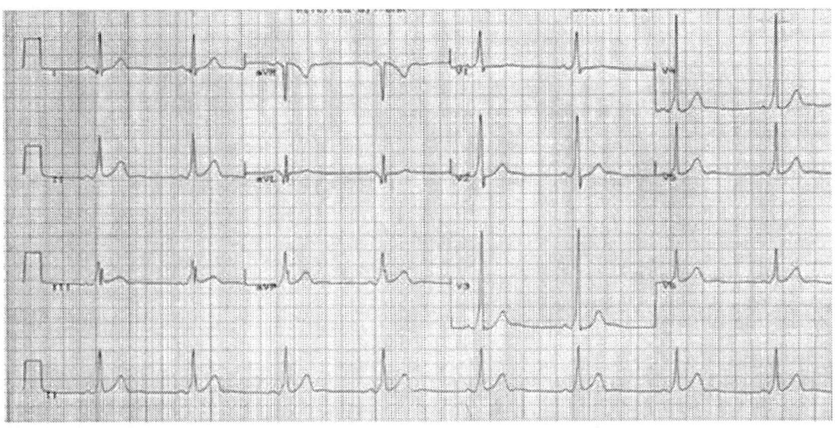

In this context of ECG recording, that has a normal heart rate, a short PR interval, a delta-wave and an early ventricular repolarization can be seen.

B/ Lown-Ganong-Levine Syndrome (LGL).

This syndrome was described in 1952 by Lown, Ganong, and Levine, forming the famous now used to describe it. It is considered a preexcitation syndrome.

We now know two types of pre-excitation syndrome:

- The Wolff-Parkinson-White or ventricular preexcitation true.
- The Lown-Ganong-Levine or accelerated atrioventricular conduction.
- Short PR alongside short QT" intervals in the same person. (Breijo's Pattern).

LGL is a disease entity that is included within the more general condition called Short PR-Interval).

Etiology • Acquired . • Congenital :

- Inherited.
- Not inherited.

The familial form is inherited, as an autosomal dominant genetic trait has been associated with the PRKAG2 gene that encodes the activated AMP protein kinase, responsible for transport and store energy from the heart. A mutation in this gene could explain the susceptibility of the heart to the crises of tachycardia. Mutation has been identified on the long arm of chromosome 7 (7q34-q36).

The Lown-Ganong-Levine may affect approximately 1 in every 50,000 people.

Several structural abnormalities have been proposed as the possible basis for LGL, including the presence of James's fibbers, Mahaim's fibbers, Brechenmacher and underdeveloped anatomic sinus node (hypoplastic).

Each of these fibbers can only be identified histologically.

Thus, unless other studies demonstrate definitive- structural or functional -abnormalities, the diagnosis of LGL remains a clinical diagnosis.

In the absence of significant structural heart disease, the mortality rate appears to be very low.

Patients may present with an acute episode of tachycardia or a history of symptoms suggestive of paroxysmal tachycardia.

In diagnosis is necessary to make:

1. A standard test for tachycardia, including an ECG to document the rhythm.
2. Serum electrolytes, calcium, magnesium levels, and levels of serum thyroid hormone-stimulating hormone (TSH). Lithemy.
3. History suggestive of recurrent paroxysms of tachycardia,
4. A Holter monitor or event recorder may be useful to document the rhythm during acute symptomatic episodes.
5. An ergometric study.
6. In rare cases, an implantable monitor for pace may be helpful.
7. Family History. (Screening).

Differential Diagnosis with Wolf-Parkinson-White

Although apparently similar, there are differences, which, in our opinion, are critical with respect to drug treatment elective. The key differences are:

- The LGL is a PR- interval shortened due to, the presence of accessory pathway, prevents the AV node but normal QRS because the accessory pathway (James fibbers) binds directly to the sinus and depolarizes the ventricles not directly, but does so by typical pathway, by the Hiss-Purkinje system.
- Not displayed "Delta waves -" in D1, aVL, V5 and V6.
- The QRS complexes tend to be narrow because there is usually no interventricular conduction disturbance.
- It is not be as frequent the association of atrial fibrillation during concomitant crisis.

Prognosis:

No studies have shown an increased risk of sudden death or reduced survival for patients meeting the criteria for the diagnosis of LGL.

Current Therapeutic Bases

Rarely, the drug medical therapy can have failures usually, but there are patients in who there is not effective (for patients who continue to have recurrent and intolerable symptoms). In such extreme cases are used:

- Radiofrequency ablation (RF)..
- The external pacemaker.
- The Implantable Cardioverter Defibrillators (ICDs).

This destroys the accessory pathway using a catheter (tube) inserted into the body to reach the heart. The success rate of this procedure ranges between 85 and 95% depending on the location of the extra or additional route.

Digoxin, verapamil and beta-blockers (other drugs commonly used to treat other types of tachycardia) can increase the frequency of episodes of tachycardia in some people with this syndrome. Beta-blockers may increase cardiac depression.

We can use drugs such as adenosine (Inpatient), and amiodarone to control or prevent episodes of tachycardia. For the control of tachycardia is usually proceed according to the severity of the implementation of vagal maneuvers carotid massage type and Valsalva maneuver (forced expiratory made with the nose and mouth closed).

Typical ECG image of the Lown-Ganong-Levine:

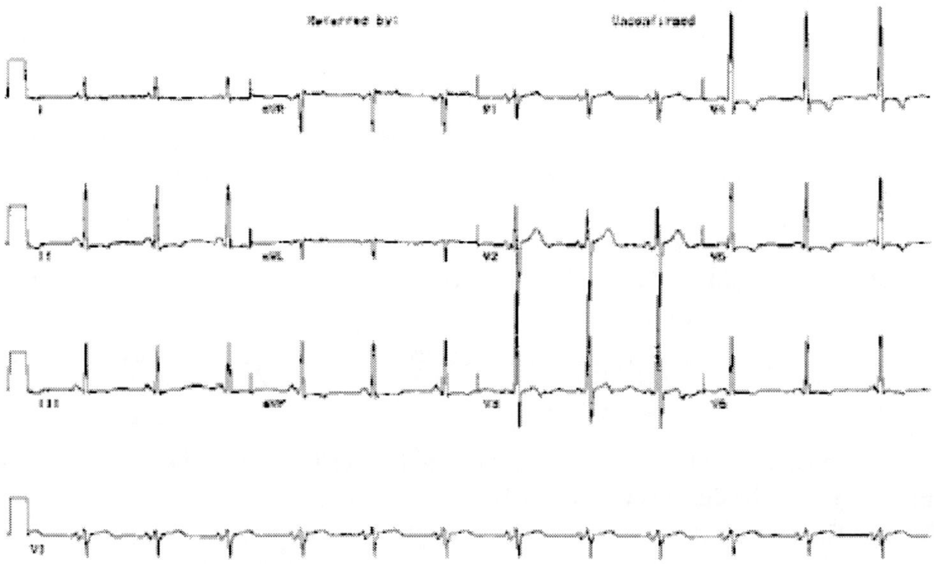

C/ Short PR-interval alongside short QT- interval on the same person. (Breijo's Pattern) [24,25,26].

In 2006, Breijo-Marquez, Pardo Ríos et al. evaluated a series of young patients, who had had, since childhood, many episodes of nocturnal palpitations, chest pain, full loss of consciousness (syncope), and which were accompanied by tonic-clonic seizures. All had been diagnosed and treated as epileptic episodes. Treatment outcomes were null. They were always considered as normal, in every cardiac studies performed absolutely.

However, all these patients had an ECG recording common:

A PR-interval lesser than 0,120 seconds with a QTc-interval equal to or lesser than 0.350 seconds.

That is, a pattern of short PR and QTc in the same person.

The correct treatment was begun (beta-blockers and, in some cases, an implantable cardio defibrillator, ICD.). Was removed all treatment from epilepsy.

The outcome to date is satisfactory.

Although we don't know, with certainty, the etiology of this pattern of ECG to date, we know that there were two important confusions:

First. - The physicians mistook to syncopal episode, with an epileptic episode.

Second. - The syncopal episodes are due to a cardiac disorder (was a cardiogenic syncope due to a cardiac electrical systole's alteration).

This ECG recording may be easily confused with a Lown-Ganong-Levine, since both have a short PR-interval. Nevertheless, in this type of ECG pattern there is also a short QTc-interval.

Unfortunately, both entities are confused with epileptic episodes too often.

Typical ECG image of the “Short PR alongside short QT” intervals in the same person. (Breijo's Pattern):

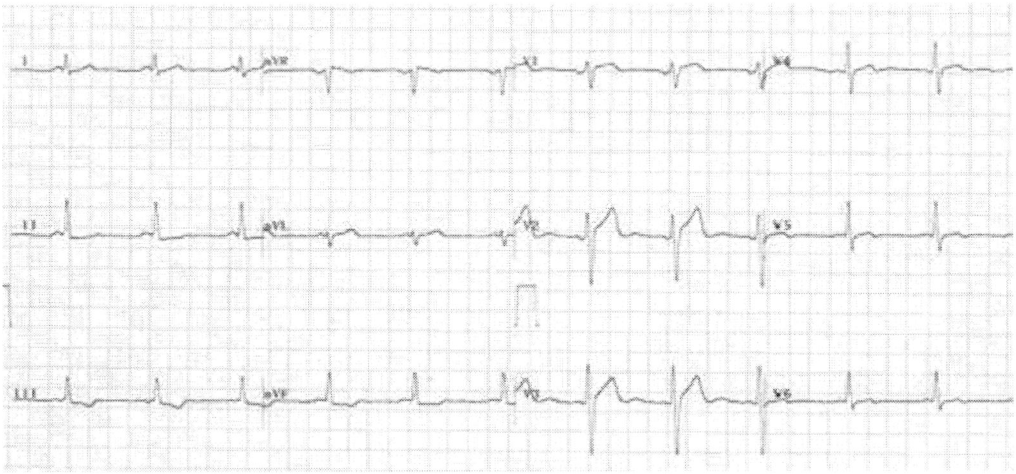

This ECG recording was the first with 12 leads that was obtained from our Hospital from Boston. MA. The patient was a 17 years-old male. We can see a shortening of the PR and QT intervals (Bazett), especially in inferior and left precordial leads. PR-interval length is lesser than 0.120 seconds and QTc length is lesser than 0.350 seconds. Patient had the symptoms exposed previously. He was also diagnosed for epileptic episodes. However, he had syncopal episodes by cardiological disturbances.

D/ MAHAIM SYNDROME.

Mahaim syndrome is characterized by:

The PR- interval with a standard length. Presence of pseudo-delta wave in the initial phase of the QRS complex because the sinus stimulus enters to AV node where physiological suffers a delay and then depolarizes the ventricles by an abnormal way: Mahaim fibers. That is:

PR-interval with a normal length.

Wide QRS complexes.

Differential diagnosis among various entities with alterations in electrical cardiac systole.

ENTITY.	PR-interval	QRS complex	QTc -interval
W.P.W	Short.	Wide (δ-wave)	Normal
L.G.L	Short	Normal	Normal
Mahaim	Normal or Short	Normal or wide	Normal
Breijo’s Pattern	Short	Normal	Short

Differential diagnosis, based on the characteristics from the different intervals and complex.

Typical ECG image of the Mahaim's Syndrome

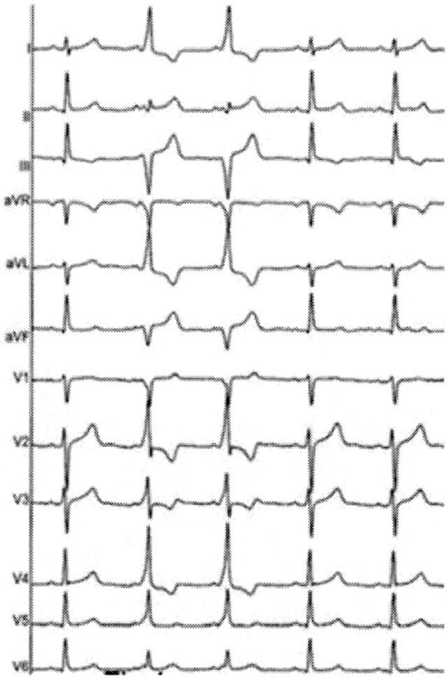

CONCLUSIONS

In any loss of consciousness, physicians should follow a protocol in order to know if a syncopal episode is or not.

1. Is This a Syncopal Episode?

On the initial presentation, the patient will not mention syncope but will use terms of common speech. They may say that they have 'fallen', 'fainted' or 'had a blackout'. A witness may state that they have fitted. It is important to assist the patient to change these general terms into more exact medical language. Start at the point where they felt normal and take them through what happened from the onset of symptoms to full recovery. When did they realize that they were not well, why was this, were there disturbances of vision, sound or smell? Did they 'fall'; do they remember the fall or just being on the floor? How quickly did they come round? How did they feel? Did they hurt themselves? Were they incontinent? Were there witnesses? Did they comment on duration, movement during unconsciousness and pallor?

2. Maybe This Is Not Syncope?

Through the history, there may be points that suggest that this is not syncope but a loss of consciousness due to another cause. The distinction between syncope and epilepsy is a difficult process and will influence the route of referral. If movements are tonic–clonic, continue for more than 15 seconds and involve the whole of a limb or limbs, this would suggest epilepsy. In comparison, irregular movements of short duration affecting only the distal parts of the limbs might suggest only cerebral hypoxia. If confusion following the attack lasts over 5–10 minutes, this may also be suggestive of an epileptic seizure. In older people with falls, a clear history may not be available. There may have been no witness of the incident and poor memory may mean events leading to the incident are unclear. Without a history of a stumble or fall then there should be a suspicion of syncope. Older people can experience syncope without realizing that it has happened. This renders the history unreliable without a witness. A full falls assessment in the patient's own environment is essential. People can collapse without losing consciousness due to other causes, such as paroxysmal benign positional vertigo and drop attacks. If there is a suspicion of collapse without loss of consciousness specialist consultation will be necessary with either an ear, nose and throat (ENT) surgeon or a neurologist. Careful definition of the term dizziness, meaning light-headedness, and giddiness, implying rotation, is important in assessment of these conditions.

3. What Has Gone Before?

Syncope needs to be put in the context of previous medical history. Is there known structural heart disease such as valvular disease or heart failure? Has the patient a past history of angina or myocardial infarction? Has the patient a past history of electrical cardiac disturbances? Has there been a previous history of blackout, especially in youth? When was this and was a diagnosis established at the time? What steps were taken? Has there been a previous diagnosis of epilepsy made? If so, how was this diagnosis made?

Not only does the sufferer need to be considered but family history also needs to be considered. Is there a history of sudden cardiac deaths? These can masquerade as unexpected deaths that have been explained away, drowning or road traffic accidents. Is there a known family history of an inherited cardiac condition?

It is also of interest to know if their parents or other family members are fainters. It has been observed that people who faint are more likely to have come from parents who were also fainters.

4. Carry Out an Examination

A clinical examination is always vital and should be comprehensive.

Starting at the pulse, this should be assessed by manual palpation for rhythm and rate. Is there a bradycardia? Is the rhythm regular? A full assessment of blood pressure both supine and erect is necessary. The blood pressure should be taken after lying for 10 minutes and then several times in the few minutes after standing, where a fall of 20 mmHg in systolic or to < 90 mmHg, or a fall in diastolic of 10 mmHg are considered abnormal and strongly suggestive of orthostatic hypotension.

The cardiovascular examination should involve a detailed auscultation of the heart to identify valvular lesions or signs of heart failure. A general neurological assessment should be undertaken, seeking gross neurological deficit. The gait should be observed to see if there are abnormalities that make falls likely.

If there is a rotational or vertiginous element in the history, a simple Dix-Hallpike assessment should be performed to check for benign paroxysmal positional vertigo. If this is positive, then assessment may be required from an ENT specialist. It is important to note that benign paroxysmal positional vertigo is not associated with loss of consciousness, and if the history is suggesting blackout, then these findings may be coincidental.

5. Undertake Investigation

Any patient presenting with syncope or a disturbed period of consciousness should undergo simple investigations. These may include blood chemistry and blood count.. An electrocardiogram (ECG) is a vital part of the assessment. A normal ECG is an excellent prognostic feature. Changes that should raise concern include:

- Bradycardia: < 40 beats per minute (except in athletes) or pauses >3 seconds
- Mobitz II, 2nd, 3rd degree atrioventricular (AV) block
- Alternating right and left bundle branch block
- Rapid supraventricular tachycardia
- Ventricular tachycardia
- Cardiac ischemia
- Myocardial infarction.
- Electrical cardiac systole's disturbances:

Wolff-Parkinson-White's syndrome, Lown-Ganong-Levine's syndrome, Mahaim's syndrome, Breijo's patterns. Long QTc-interval. Short QTc-interval. Brugada' syndrome. ECG patterns mixed. Unless the attacks are happening on a daily or very frequent basis, 24-hour ECG monitoring reveals nothing. A normal 24-hour ECG without symptoms does not further the diagnosis at all and should not be seen as reassuring.

6. What Is the Prognosis?

Patients presenting with any problem want to know if they will come to any harm. This is such a complex area that it can be difficult to give assurances. However, in young healthy people with a normal ECG and a suggestion of neurally mediated (vagal) or rarely orthostatic hypotension the prognosis is excellent. The presence of structural heart disease, independent of the cause of syncope, reflects a poor prognosis.

SOME EXAMPLES OF ECG RECORDING WITH DIFFERENT PATTERNS THAT PRODUCE SYNCOPAL EPISODES. [27,28].

1°. Wolff-Parkinson-White and Prolonged"Q-T" Patterns in the Same Electrocardiographic Record

Wolff-Parkinson-White syndrome (WPWS) is a congenital heart disease (PRKAG2. Genetic map 7q36) characterised by a premature ventricular depolarisation caused by an abnormal atrioventricular accessory pathway known as Kent's bundle. Prolonged QT syndrome (PQTS) consists of an abnormal prolongation of the QT interval on the ECG, which can be both inherited and acquired. This anomaly is known to favour the occurrence of malign cardiac arrhythmias, above all polymorphic ventricular tachycardia, ventricular fibrillation and "torsade de pointes".

When taken separately, both syndromes have little incidence, which leads us to expect this incidence to be even lower when they are found on the same electrocardiogram. Incidentally, the current medical literature contains no publications on this topic. This clinical case aims to establish the existence of an electrocardiographic pattern characterised by WPW and a PQTS pattern on an ECG record. With a high susceptibility to crisis of tachycardia, especially at night, several episodes of syncope, even cardiac arrest.

The patient is a 24 old- years man. Since childhood, he has suffered from more than four tachycardia attacks, three documented syncope episodes, as well as two cardiac arrests recovered, for which he was treated with electric discharges. Afterwards, he was treated with radiofrequency ablation of Kent's bundle, with permanent positive results so far.

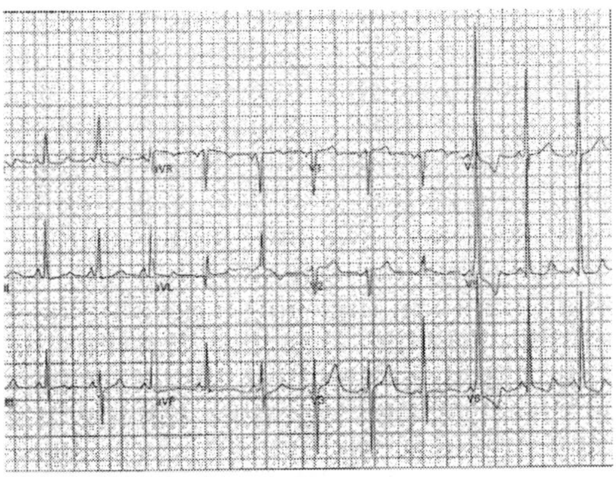

We can see a typical ECG recording of an intermittent WPW and a Long QT-interval together in a patient with several syncopal episodes and a recovered cardiac arrest.

2°. ECG patterns with short PR interval together to a long QT (A) and first-degree AV block alongside a long QT (B: Increased of cardiac electrical systole). [25,27,29].

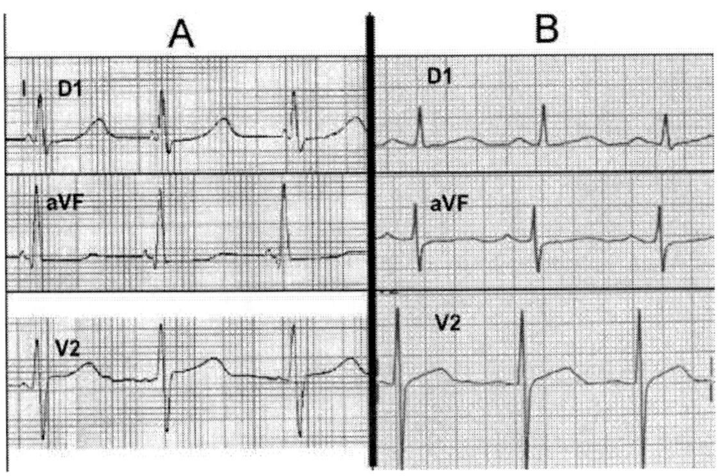

In this exposition, we present the ECG record of two patients with an obvious diversity and variability of alterations in the electrical system of heart. These electrical cardiac disturbances could explain completely the symptomatology from patients: nocturnal palpitations, several syncopal episodes. In Figure A, we can see the presence of a short PR-interval together to a Long QT-interval. In Figure B, we can also see an ECG recording with a Long PR-interval alongside a Long QT-interval.

ACKNOWLEDGMENTS

Lourdes and Alejandro Breijo, for their unselfishness collaboration from Miami. Florida. USA.

REFERENCES

[1] Brignole M, Alboni P, Benditt DG, Bergfeldt L, Blanc JJ, Bloch Thomsen PE, et al; Task Force on Syncope, European Society of Cardiology. Guidelines on management (diagnosis and treatment) of syncope —update 2004. *Europace*. 2004;6:467-537.

[2] Thijs RD, Benditt DG, Mathias CJ, Schondorf R, Sutton R, Wieling W, et al. Unconscious confusion —a literature search for definitions of syncope and related disorders. *Clin. Auton Res*. 2005;15:35-9.

[3] Soteriades ES, Evans JC, Larson MG, Chen MH, Chen L, Benjamin EJ, Levy D. Incidence and prognosis of syncope. *N. Engl. J. Med.* 2002;347:878-85.

[4] Strickberger SA, Benson DW, Biaggioni I, Callans DJ, Cohen MI, Ellenbogen KA, et al; American Heart Association Councils on Clinical Cardiology, Cardiovascular Nursing, Cardiovascular Disease in the Young, and Stroke; Quality of Care and Outcomes Research Interdisciplinary Working Group; American College of Cardiology Foundation; Heart Rhythm Society. AHA/ACCF scientific statement on the evaluation of syncope. *J. Am. Coll Cardiol*. 2006;47:473-84.

[5] Hoefnagels WA, Padberg GW, Overweg J, Van der Velde EA, Roos RA. Transient loss of consciousness: the value of the history for distinguishing seizure from syncope. *J. Neurol.* 1991;238:39-43.

[6] Stephenson J. Fits and Faints. Oxford: Blackwell Scientific Publications; 1990. p. 41-57.

[7] Van Dijk JG, Sheldon R. Is there any point to vasovagal syncope? *Clin. Auton Res.* 2008;18:167-9.

[8] Tea SH, Mansourati J, L'Heveder G, Mabin D, Blanc JJ. New insights into the pathophysiology of carotid sinus syndrome. *Circulation*. 1996;93:1411-6.

[9] Alboni P, Alboni M, Bertorelle G. The origin of vasovagal syncope: to protect the heart or to escape predation? *Clin. Auton Res.* 2008;18:170-8.

[10] Mathias CJ, Mallipeddi R, Bleasdale-Barr K. Symptoms associated with orthostatic hypotension in pure autonomic failure and multiple system atrophy. *J. Neurol.* 1999;246:893-8.

[11] Naschitz J, Rosner I. Orthostatic hypotension: framework of the syndrome. *Postgrad. Med. J.* 2007;83:568-74.
[12] Consensus statement on the definition of orthostatic hypotension, pure autonomic failure, and multiple system atrophy. *J. Neurol. Sci.* 1996;144:218-9.
[13] Wieling W, Krediet P, Van Dijk N, Linzer M, Tschakovsky M. Initial orthostatic hypotension: review of a forgotten condition. *Clin. Sci.* (Lond). 2007;112:157-65.
[14] Podoleanu C, Maggi R, Brignole M, Croci F, Incze A, Solano A, et al. Lower limb and abdominal compression bandages prevent progressive orthostatic hypotension in the elderly. A randomized placebo-controlled study. *J. Am. Coll Cardiol.* 2006;48:1425-32.
[15] Gibbons CH, Freeman R. Delayed orthostatic hypotension: a frequent cause of orthostatic intolerance. *Neurology.* 2006;67:28-32.
[16] Verheyden B, Gisolf J, Beckers F, Karemaker JM, Wesseling KH, Aubert A, et al. Impact of age on the vasovagal response provoked by sublingual nitroglycerine in routine tilt testing. *Clin. Sci.* (Lond). 2007;113:329-37.
[17] Grubb BP, Kosinski DJ, Boehm K, Kip K. The postural orthostatic tachycardia syndrome: a neurocardiogenic variant identified during head-up tilt table testing. *Pacing. Clin. Electrophysiol.* 1997;20:2205-12.
[18] Leitch JW, Klein GJ, Yee R, Leather RA, Kim YH. Syncope associated with supraventricular tachycardia: an expression of tachycardia or vasomotor response? *Circulation.* 1992;85:1064-71.
[19] Brignole M, Gianfranchi L, Menozzi C, Raviele A, Oddone D, Lolli G, et al. Role of autonomic reflexes in syncope associated with paroxysmal atrial fibrillation. *J. Am. Coll Cardiol.* 1993;22:1123-9.
[20] Ebert SN, Liu XK, Woosley RL. Female gender as a risk factor for drug-induced cardiac arrhythmias: evaluation of clinical and experimental evidence. *J. Womens Health.* 1998;7:547-57.
[21] Zareba W, Moss AJ, Le Cessie S, Locati EH, Robinson JL, Hall WJ, Andrews ML. Risk of cardiac events in family members of patients with Long QT syndrome. *J. Am. Coll Cardiol.* 1995;26:1685-91.
[22] Lombroso CT, Lerman P. Breathholding spells (cyanotic and pallid infantile syncope). *Pediatrics.* 1967;39:563-81.
[23] Wieling W, Ganzeboom KS, Saul JP. Reflex syncope in children and adolescents. *Heart.* 2004;90:1094-100.
[24] Breijo-Marquez, F.R. Decrease of electrical cardiac systole. *Int. J. Cardiol.* 2008;126:e36-8.
[25] Breijo Marquez, FR. Pardo Ríos, M. Variability and diversity of electrical cardiac systole. *BMJ Case Reports*. 2009.
[26] Breijo Marquez, F. Pardo Ríos, M. Alcaraz Baños, M. Association of Short PR Interval, Long QT Interval, and Sudden Cardiac Death in a Young Male. *Rev. Esp. Cardiol.*2010; 63 :362-4.
[27] Presence of a critical stenosis in left anterior descending coronary artery alongside a short “P-R” and “Q-T” pattern, in the same electrocardiographic record. *J. Electrocardiology*. 2010; 422-424.
[28] Breijo-Marquez, FR. Pardo Ríos, M. Wolff-Parkinson-White and Prolonged “Q-T” Patterns in the Same Electrocardiographic Record. *J. Clinic Experiment Cardiol* 2:118.

[29] Breijo-Marquez, Fr. Pardo Ríos, M. Sudden cardiac death in a young adult with diagnosed with Tietze syndrome. *Rev. Esp. Dolor*. 2010. 56: 321-6.

In: Comas and Syncope: Causes, Prevention and Treatment
Editors: Eduardo Silva and Gabriel Cruz.
ISBN 978-1-62100-603-9

Chapter 4

SYNCOPE: PREVENTION, RECOGNITION, AND TREATMENT

Chandrasekar Palaniswamy, Wilbert S. Aronow*, Nikhil Agrawal, Nivas Balasubramaniyam and Sachin Sule

Department of Medicine, Division of Cardiology,
New York Medical College, Valhalla, NY, US

ABSTRACT

Syncope, defined as a sudden, transient loss of consciousness with spontaneous recovery, may be the presenting symptom for conditions ranging from benign illnesses to life-threatening disorders. Common causes of syncope include vasovagal syncope, situational syncope, carotid sinus hypersensitivity, orthostatic hypotension, structural heart diseases, left and right sided obstructive cardiac lesions, and cardiac arrhythmias. History often provides valuable clues about the underlying etiology of syncope. Admission decisions in the emergency department are guided by various risk prediction scores. Evaluation of a patient with syncope involves a large battery of diagnostic tests that include a 12-lead electrocardiogram, Holter monitoring, an echocardiogram, tilt table testing, ischemia evaluation, electrophysiologic studies, and other imaging tests. Despite the availability of these advanced diagnostic tests, a significant proportion of patients with syncope remain undiagnosed. Therapy includes various non-pharmacological measures such as life style modifications, adequate fluid and salt intake, tilt training exercises, and pharmacotherapy with fludrocortisone, midodrine, beta adrenergic blockers, vagolytics, or serotonin reuptake inhibitors. Cardiac pacing may be beneficial in a selected subgroup of patients.

* Address for correspondence: Wilbert S. Aronow, MD, FACC, FAHA Cardiology Division, New York Medical College Macy Pavilion, Room 138 Valhalla, NY, 10595, USA Telephone: 914-493-5311 Fax: 914-235-6274 E-mail: wsaronow@aol.com

INTRODUCTION

Syncope is a transient loss of consciousness and postural tone due to transient global cerebral hypoperfusion characterized by rapid onset, short duration, and spontaneous recovery. Syncope may occur suddenly, without warning, or may be preceded by prodromal symptoms that include lightheadedness, dizziness, a feeling of warmth, diaphoresis, nausea, visual blurring, or tunnel vision. Although syncope may be a presenting symptom for conditions ranging from benign illnesses to life-threatening disorders, cerebral hypoperfusion is the common final pathway for the majority of episodes. The term syncope derives from the Greek word *synkoptein,* meaning "to cut short".

Syncope accounts for 1% to 3% of emergency department visits and up to 6% of hospital admissions. [1, 2] The total annual costs of syncope-related admissions exceed $2 billion. [3] Data from the Framingham Heart Study (from 1971 to 1998) indicate that the incidence of a first report of syncope was 6.2 per 1000 person-years. [4] The age distribution of syncopal episodes is bimodal, with the first peak between the age of 10 and 30 years, and a second peak in patients greater than 65 years of age. The study also concluded that persons with cardiac syncope are at increased risk for death from any cause and cardiovascular events, and persons with syncope of unknown cause are at increased risk for death from any cause.

CAUSES OF SYNCOPE

Syncope is usually due to one of two general mechanisms: disorders of vascular tone or blood volume, or cardiovascular disorders including obstructive lesions and cardiac arrhythmias. Common causes of syncope are listed in table 1.

Disorders of Vascular Tone

Neurocardiogenic (Vasovagal and Vasodepressor) Syncope

This includes the common "faint" that may be experienced by normal persons, and accounts for approximately half of all episodes of syncope. This is frequently recurrent and commonly precipitated by a warm, crowded place, following prolonged standing, hunger, extreme fatigue, and experiencing fear or pain. Episodes are often preceeded by prodromal signs and symptoms were like nausea, diaphoresis, lightheadedness, visual blurring and pallor that offer warning of an impending episode of syncope. Sometimes the patient remains partly aware of the surroundings, or there may be complete unresponsiveness. A few clonic jerks of the limbs and face may occur, but this should not be confused with a seizure. In contrast to a seizure, sphincter control is usually maintained in a syncopal event. Although vasovagal syncope is usually benign, this can sometimes be associated with prolonged asystole and hypotension, with possible hypoxic-ischemic injury. The proposed mechanism of neurocardiogenic syncope is illustrated in figure 1. Although the principal mechanism involves myocardial mechanoreceptors, occurrence of neurocardiogenic syncope in patients with transplanted (denervated) hearts suggests additional central mechanisms like serotonin surge that may contribute to the sympathetic withdrawal [5].

Table 1. Causes of Syncope

Vascular
Reflex
1. Neurocardiogenic (Vasovagal/Vasodepressor)
2. Situational (Cough, micturition, defecation etc)
3. Carotid Sinus Hypersensitivity
4. Glossopharyngeal Nerve stimulation

Orthostatic
Volume Depletion
Autonomic Insufficiency
Pharmacological Agents
Venous pooling

Cardiac
A) Structural / Obstructive
1. Aortic Stenosis
2. Hypertrophic cardiomyopathy
3. Mitral stenosis
5. Cardiac tamponade
6. Myocardial infarction
7. Pulmonary embolism
8. Pulmonary hypertension

B) Arrhythmias
1. Bradyarrhythmias (sick sinus syndrome, atrioventricular block)
2. Tachyarrhythmias (Supraventricular tachycardias, Atrial fibrillation with rapid ventricular rate, Atrial flutter with 1:1 conduction, Ventricular tachycardia -structural heart disease, inherited long QT syndrome, drug induced long QT syndrome, Brugada syndrome, arrhythmogenic right ventricular cardiomyopathy)
3. Pacemaker Malfunction (pacemaker mediated tachycardia, pacemaker syndrome)

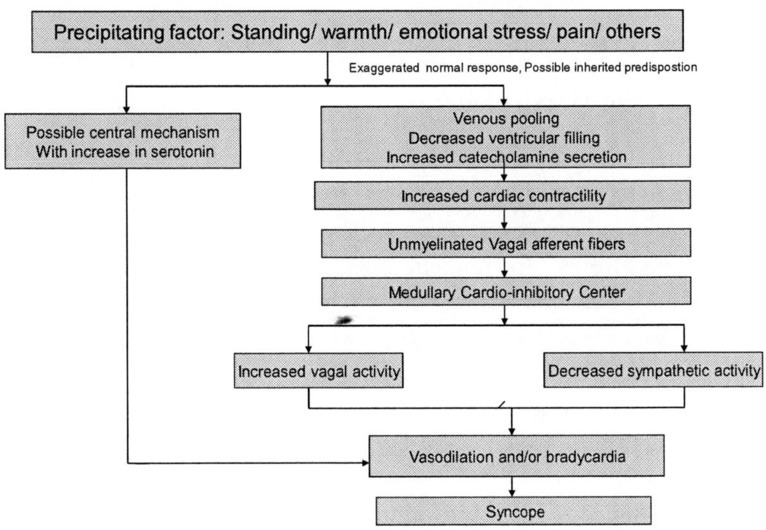

Figure 1. Mechanism of Neurocardiogenic Syncope.

Situational Syncope

In some individuals, increased intrathoracic pressure that occurs with cough, micturition, and defecation may contribute to hypotension and syncope by decreasing venous return. Cough syncope typically occurs in men with chronic obstructive airway disease, micturition syncope in men with prostatic hypertrophy, and defecation syncope in older individuals with chronic constipation. Syncope usually occurs at during or immediately after the precipitating factor in these patients.

Carotid Sinus Hypersensitivity

Carotid sinus hypersensitivity occurs predominantly in older men who present with either syncope or falls. [6] This is precipitated by pressure on the carotid sinus baroreceptors, which are located just cephalad to the bifurcation of the common carotid artery. This typically occurs in the setting of shaving, a tight collar, or turning the head to one side. The proposed mechanism of carotid sinus hypersensitivity is illustrated in figure 2. Activation of carotid sinus baroreceptors in hypersensitive individuals results in the following: sinus arrest or AV block (a cardioinhibitory response), vasodilatation (a vasodepressor response), or both (a mixed response).

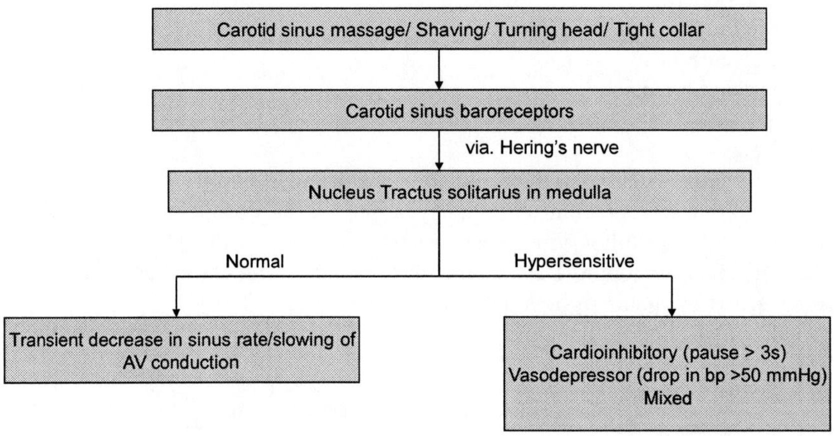

Figure 2. Mechanism of carotid hypersensitivity.

Orthostatic Hypotension

Orthostatic intolerance can result from hypovolemia or from disturbances in vascular control. Sudden rising from a recumbent position or standing quietly are precipitating circumstances. Postural syncope may occur in otherwise normal persons with defective postural reflexes. In some cases, the autonomic nervous system may fail entirely, resulting in severe orthostatic hypotension (a fall of ≥20 mm Hg systolic or ≥10 mm Hg diastolic blood pressure within 2 minutes of standing). [7] The primary forms of this disorder include pure autonomic failure, multiple system atrophy, and Parkinsonian dysautonomia. Postural orthostatic tachycardia syndrome (POTS) is a less severe form of autonomic insufficiency in which heart rate increases excessively in response to upright posture without a drop in blood pressure. [8] It is of 2 types: peripheral dysautonomic form and central hyperadrenergic form. Patients with dysautonomic POTS, the commoner form, often complain of palpitations, fatigue, lightheadedness, dizziness, near syncope, and syncope that is related to upright posture. Some patients with POTS may progress to pure autonomic failure. In contrast, patients with central hyperadrenergic POTS often complain tremors, hyperhidrosis, diarrhea, and anxiety that is related to high levels of serum catecholamines that are released in upright posture. Orthostatic hypotension may also be secondary to the effects of pharmacologic therapies like diuretics, nitrates, vasoactive medications, chemotherapeutic agents, and disease states like diabetes, amyloidosis etc. In a prospective analysis of 499 long-term health care facility residents, 118 (24%) had a maximal decrease in postprandial systolic blood pressure of > or = 20 mm Hg. The mean maximal decrease in postprandial systolic blood pressure was 24 +/- 5 mm Hg in residents with syncope in the prior 6 months and 14 +/- 5 mm Hg in residents without syncope ($P < 0.0001$). [9] At long-term follow-up, postprandial systolic hypotension in these patients was associated with a higher incidence of falls, syncope, new coronary events, new stroke, and total mortality [10].

Cardiac Causes

Syncope can occur from either severe obstruction at the level of right or left heart, or from disturbances of cardiac rhythm that lead to reduced cardiac output.

Structural Heart Diseases

Syncope may be the presenting symptom in patients with left-sided obstructive lesions, including aortic stenosis and hypertrophic cardiomyopathy. Occurrence of syncope in these patients usually suggests that the obstruction is severe. Other anatomic causes of syncope include massive pulmonary embolus, severe pulmonary hypertension, atrial myxoma and cardiac tamponade.

Arrhythmic Causes

Tachyarrhythmias and bradyarrhythmias are a common cause of syncope, especially in patients with preexisting cardiac disease. Ventricular tachycardia (VT) is the most common tachyarrhythmia that can cause syncope. Supraventricular tachycardias (SVT) can also cause syncope when the ventricular rate is very high and impairs diastolic filling. Bradyarrhythmias that can result in syncope include sick sinus syndrome and atrioventricular (AV) block. An important aspect of sinus node dysfunction that may easily be missed is evidence of chronotropic incompetence. Patients may have reasonable sinus rates at rest, but may be unable to mount a sufficient heart rate response during exercise. Chronotropic incompetence is generally defined as an inability to achieve a heart rate of 70-85% of the age predicted maximum (220 - age) with exercise.

Risk Prediction Algorithms

Although many clinical algorithms have been published for predicting adverse outcomes in patients hospitalized with syncope, none have been widely implemented. The commonly used ones include Osservatorio Epidemiologico sulla Sincope nel Lazio (OESIL) risk score and San Francisco syncope rule (SFSR) (Table 2). [11] For the OESIL score, four patient characteristics among patients presenting with syncope were associated with adverse outcome: age greater than 65 years, a clinical history of cardiovascular disease, syncope without prodromal symptoms, and abnormal ECG findings. The presence of each characteristic scored one point. Patients were considered to have cardiovascular disease if they had any form of structural heart disease, prior or clinical evidence of congestive heart failure, peripheral arterial disease, or stroke or transient ischemic attack. ECG findings were considered abnormal if there were rhythm abnormalities, atrio-ventricular or intraventricular conduction disorders, left or right ventricular hypertrophy, left-axis deviation, previous myocardial infarction, or ST-segment and T-wave abnormalities consistent with or possibly related to myocardial ischemia. In the OESIL study, one-year mortality was shown to increase with an increasing number of risk factors, from 0% for no risk factors to 57% for four risk factors. [12] In the SFSR study, a high-risk patient with syncope was defined as having any of the following five risk factors: abnormal ECG findings, a complaint of dyspnea, hematocrit <30%, systolic BP <90 mm Hg, or a history of congestive heart failure. The presence of any of the five risk factors was predictive of short-term serious outcomes (within 7 days of the initial visit), with a 96% sensitivity and 62% specificity. [13] Both the OESIL score and SFSR were able to identify an increased probability of medium-term (3 months) serious outcome in patients with syncope. [14] In a meta-analysis to assess the methodological quality and prognostic accuracy of clinical decision rules in emergency department patients, the sensitivity and specificity were 86% and 49% for SFSR, and 95% and 31% for OESIL score respectively [15].

Table 2. Risk prediction scores in evaluation of syncope

San Francisco Syncope Rule (SFSR)
Abnormal electrocardiogram
History of congestive heart failure
History of dyspnea
Hematocrit <30%
Systolic blood pressure < 90 mmHg
Osservatorio Epidemiologico sulla Sincope nel Lazio (OESIL) Score
Abnormal electrocardiogram
History of cardiovascular disease
Absence of prodrome
Age >65years
Evaluation of Guidelines in Syncope Study (EGSYS) Score
1. Palpitation before syncope(+4)
2. Abnormal EKG or heart disease(+3)
3. Syncope during effort(+3)
4. Syncope while supine(+2)
5. Autonomic prodrome (-1)
6. Predisposing or precipitating factor(-1)

In the Evaluation of Guidelines in Syncope Study (EGSYS), a prospective study of 516 patients with unexplained syncope, mean age of 63 years, abnormal ECG findings and/or heart disease, palpitations before syncope, syncope during effort or in the supine position, absence of prodromal symptoms, and absence of precipitating factors were found to be predictors of cardiac syncope. A score from+4 to −1 was assigned to each variable according to the magnitude of regression coefficient. A score of 3 identified cardiac syncope, with a sensitivity of 92% to 95% and a specificity of 61% to 69% in the derivation and validation cohorts [16-18].

These scores are primarily designed to predict patients at risk for short-term serious outcomes and help guide admission decisions. However, these scores should not be relied on for the prediction of mortality or readmission on long term follow-up. While the short term outcomes are predominantly related to the underlying etiology and the risk factors mentioned above, the long-term mortality is related to the comorbidities. We had investigated the etiologies of syncope and risk factors for mortality and rehospitalization for syncope at 27-month follow-up in 325 consecutive patients, mean age 66 years, hospitalized for syncope. [19] The causes of syncope were diagnosed in 241 patients (74%). Of 325 patients, 13 (4%) were rehospitalized for syncope and 38 (12%) died. Stepwise logistic regression analysis showed that significant independent prognostic factors for rehospitalization for syncope were diabetes (odds ratio [OR], 5.7; 95% confidence interval [CI], 1.6–20.4), atrial fibrillation (OR, 4.0; 95% CI, 1.0–15.6), and smoking (OR, 4.6; 95% CI, 1.3–16.8). Significant independent prognostic factors for time to mortality were diabetes (hazard ratio [HR], 2.7; 95% CI, 1.4–5.2), coronary artery bypass graft surgery (HR, 2.9; 95% CI, 1.3–6.5), malignancy history (HR, 2.5; 95% CI, 1.2–5.2), narcotics use (HR, 4.0; 95% CI, 1.7–9.8), smoking (HR, 2.8; 95% CI, 1.4–5.5), atrial fibrillation (HR, 2.4; 95% CI, 1.0–5.4), and volume depletion (HR, 2.8; 95% CI,1.4–5.8) [19].

Measurement of natriuretic peptide levels may also be helpful in differentiating cardiac from non-cardiac syncope. Further studies are needed to define their role of in the diagnostic algorithm of syncope [20,21].

Diagnosis

A patient who has lost consciousness should be placed in a position that maximizes cerebral blood flow, offers protection from trauma, and secures the airway. Whenever possible, the patient should be placed supine with the head turned to the side to prevent aspiration and the tongue from blocking the airway. If cervical spine injury is suspected, neck should be immobilized. Clothing that fits tightly around the neck or waist should be loosened. The physician should quickly rule out those causes that constitute a therapeutic emergency, including hypoglycemia, massive internal hemorrhage, myocardial infarction, or cardiac arrhythmias. The nature of the events immediately prior to, during, and after an episode of syncope often provides valuable etiologic clues. Syncope in the supine position is unlikely to be of vasovagal etiology. Syncope due to carotid sinus hypersensitivity may occur when the individual is wearing a shirt with a tight collar, turning the head, or manipulating the neck. The patient's medications must be noted, including nonprescription drugs or health store supplements, with particular attention to recent changes. In elderly persons, complete heart block or a tachyarrhythmia should be suspected, even though all findings are negative when the patient is seen. The physical examination should include evaluation of heart rate and blood pressure in the supine, sitting, and standing positions. The choice of diagnostic tests should be guided by the history and the physical examination (Table 3).

Table 3. Diagnostic Tests in evaluation of syncope

Test	Comments
Electrocardiogram	Initial standard part of evaluation
Tilt-table testing	Unknown etiology with multiple attacks or associated with injury. Known etiology but diagnosis of neurocardiogenic syncope can change management.
Echocardiogram	Evaluate structural etiology
Holter monitoring	Recurrent frequent syncope
External/ Implantable loop recorders	Recurrent infrequent syncope
Pacemaker/ Defibrillator interrogation	In patients with pacemaker/ defibrillator in situ
Electrophysiologic studies	See table 4
Stress test/ Cardiac catheterization	If exertion induced/ associated with angina/ high risk for CAD
Spiral CT of chest/ Ventilation-Perfusion scan	If Pulmonary embolism is suspected
Coronary CT angiography	Coronary artery anomalies
Cardiac Magnetic Resonance Imaging	Evaluate for arrhythmogenic right ventricular cardiomyopathy

In-patient evaluation of a patient with syncope frequently involves a large battery of diagnostic tests with substantial health care costs. Implementation of a “Syncope Unit” allows an improved management of patients with syncope with reduced economic costs. [22] Even after a thorough assessment, it may not be possible to identify an etiology for syncope. On the other hand, patients often have multiple comorbidities and as a result they may have several equally probable causes of syncope. The physician must be judicious in deciding whether an observed abnormality is the cause of syncope in a given individual. Measurements of serum electrolytes, glucose, and the hematocrit are usually indicated. Cardiac enzymes should be evaluated if myocardial ischemia is suspected. Blood and urine toxicology screens may reveal the presence of alcohol or other drugs.

Electrocardiographic Monitoring

Correlation of the cardiac rhythm with symptoms is the best method for determining the etiology of complaints such as dizziness or syncope. Although the surface electrocardiogram (ECG) is unlikely to provide a definitive diagnosis, it may provide clues to the cause of syncope and should be performed in almost all patients. The presence of conduction abnormalities (PR prolongation and bundle branch block) suggests a bradyarrhythmia, whereas pathologic Q waves or prolongation of the QT interval suggests a ventricular tachyarrhythmia. Inpatients should undergo continuous electrocardiographic monitoring; outpatients should wear a holter monitor for 24 to 48 hours. Continuous electrocardiographic monitoring may establish the cause of syncope in as many as 15% of patients. In patients with recurrent symptoms who remain alert enough during an event, event monitors may be very useful. This is not useful for a patient who has no warning of an event. In this scenario, external loop recorders may be useful. [23, 24] For patients in whom the etiology of syncope remains unclear after ECG monitoring, Implantable Loop Recorders (ILR) may be used for long-term monitoring. [25, 26] These devices can be implanted subcutaneously in the right parasternal, subcostal, and axillary regions of the chest. Unlike a pacemaker, they do not require any leads in the heart. The device records a bipolar ECG. The battery life of currently available ILRs is approximately 3 years. Data storage can be activated in two fashions. When the patient suffers a syncopal event, the memory can be frozen using a hand-held activator given to the patient at the time of implant. The device also has an automatic detection function capable of recording rhythm strips of high and low heart rate episodes as well as pauses, in addition to five manually activated tracings. The stored events are then downloaded into a standard pacemaker programmer and analyzed. The device has been shown to be quite valuable in the diagnosis of syncope of unknown origin. Utility of signal-averaged ECG, microvolt T-wave alternans and heart rate variability have been studied, but are not routinely used in evaluation of patients with syncope [27-29].

Head-Up Tilt Test

Tilt table testing is indicated for recurrent syncope, a single syncopal episode that caused injury, or a single syncopal event in a "high-risk" setting (pilot, commercial vehicle driver etc). In susceptible patients, upright tilt at an angle between 60° (commonly used) and 80° for 30–60 min induces a vasovagal episode. [30] Abnormal tilt-table responses can be classified into one of the following five groups. [31] The first is a classic neurocardiogenic response, which is characterized by an abrupt drop in blood pressure. When accompanied by a significant drop in heart rate, this is referred to as a vasovagal response; in the absence of a decrease in heart rate, a vasodepressor response. A second pattern referred to as a dysautonomic (or delayed orthostatic), is characterized by a gradual progressive fall in blood pressure with relatively little change in heart rate. This pattern is often noted in the autonomic failure syndromes. The third pattern, termed a postural tachycardic response, is associated with a >30 beats/min increase in heart rate (or a heart rate of >120 beats/min) during the first 10 minutes of the baseline tilt. The fourth pattern is called cerebral syncope. These patients experience syncope in the absence of systematic hypotension concomitant with intense cerebral vasoconstriction (determined by transcranial Doppler), as well as cerebral hypoxia (as determined by electroencephalogram). The fifth response pattern is psychogenic. These patients are noted to experience loss of consciousness in the absence of any observable change in heart rate, blood pressure, electroencephalogram or transcranial Doppler recording. These patients are often found to suffer from psychiatric disorders that range from conversion reactions to severe depression. The protocol can be shortened if upright tilt is combined with administration of drugs that cause venous pooling or increase adrenergic stimulation (isoproterenol, nitroglycerin, edrophonium, or adenosine). A baseline tilt duration of at least 30-45 minutes should be employed prior to any pharmacological intervention. [32] Isoproterenol infusion increases the yield of tilt table testing considerably, but it may produce a higher incidence of false-positive responses than tilting without isoproterenol. Sublingual nitroglycerin tablet and intravenous adenosine have also been employed as alternative forms of pharmacologic provocation. Tilt table testing is highly effective for detecting patients with vasovagal syncope. The specificity of tilt-table testing has been reported to be near 90%, but it is lower when pharmacologic provocation is employed. The reported sensitivity of the test ranges between 20 and 74%, the variability due to differences in populations studied, techniques used, and the absence of a true "gold standard" against which to compare test results. The accuracy of serial tilt table testing in predicting efficacy of therapy is not well established.

Echocardiogram, Stress Tests, and Cardiac Catheterization

The echocardiogram with Doppler examination detects valvular, myocardial, and pericardial abnormalities. An echocardiogram should be performed whenever the diagnosis of structural heart disease is considered to be significantly likely. The echocardiogram is the test of choice for the diagnosis of aortic stenosis, hypertrophic cardiomyopathy and atrial myxoma. Patients who are at risk for coronary artery disease should undergo a stress test. In addition to its use to detect myocardial ischemia, an exercise stress test should be performed when exertional arrhythmias are suspected and ambulatory monitoring has failed to document arrhythmia. In patients who are at high risk for or who are known to have coronary artery disease, cardiac catheterization may be indicated. In some patients, cardiac catheterization may be necessary to diagnose the presence or severity of coronary artery disease or valvular abnormalities.

Electrophysiologic (EP) Studies

The indications of EP study in evaluation of syncope are summarized in table 4. If the patient has a normal ECG and no structural heart disease, the yield from EP study is so low that it will rarely be useful. On the other hand, an older patient with known CAD and syncope is much more likely to have an arrhythmia such as VT as the etiology. If there is no other clear cause for syncope, an EP study is indicated to determine whether there is inducible VT in order to institute appropriate treatment. In a patient undergoing an EP study for syncope, tests of sinus and AV nodal function are performed. [33] The sinus node recovery time (SNRT) is obtained by pacing the atrium rapidly for 30-60 seconds followed by abruptly terminating the pacing and measuring the time it takes for the first spontaneous sinus beat to occur. The corrected SNRT is calculated by subtracting the sinus cycle length from the SNRT. The normal value is <525 msec. EP testing of the AV node involves evaluation of the atrial pacing rate at which Wenckebach AV block occurs, determination of the refractoriness of the AV node, and examination for any infranodal block. In addition, the patient must be evaluated for any inducible tachyarrhythmias, both supraventricular and ventricular, which might have precipitated syncope. EP testing for syncope is much more likely to reveal an abnormality when the patient has a tachyarrhythmia as the etiology for syncope [34].

Table 4. Indications for Electrophysiologic Testing in the Evaluation of Patients with Syncope

In patients with ischemic heart disease when the initial evaluation suggests an arrhythmic cause and there is no established indication for a defibrillator
In patients with bundle branch block when noninvasive tests do not establish a diagnosis
In patients with syncope of abrupt onset or preceded by palpitations when noninvasive tests do not establish a diagnosis
In selected patients with syncope and Brugada syndrome, arrhythmogenic right ventricular cardiomyopathy and hypertrophic cardiomyopathy
In patients with high-risk occupations, in whom every effort to exclude a cardiovascular cause of syncope is warranted

Other Imaging Tests

Cardiac magnetic resonance imaging (MRI) provides an alternative noninvasive modality that may be useful for patients in whom diagnostic-quality echocardiographic images cannot be obtained. This test is also indicated for patients suspected of having arrhythmogenic right ventricular dysplasia or right ventricular outflow tract ventricular tachycardia. Both are associated with right ventricular structural abnormalities that are better visualized on MRI than by echocardiogram. Ultrafast CT scan, ventilation-perfusion scan, or pulmonary angiography is indicated in patients in whom syncope may be due to pulmonary embolus.

Despite the use of EP studies and head-up tilt testing, a significant proportion of patients who present for evaluation of syncope remain undiagnosed. Such patients have a fairly good prognosis, although not that of a normal population. Those with underlying cardiac disease are at greatest risk, and those without known cardiovascular disease have minimal risk of premature death.

Treatment

Non-Pharmacological Treatments

Treatment of syncope mainly depends on the cause. Regardless of the cause certain general measures need to be adopted. Patients with neurocardiogenic syncope should be counseled to avoid circumstances that could precipitate an episode of syncope like alcohol, extreme fatigue, hot or crowded place, severe pain, hunger and other stressful situations. Patients with frequent episodes should be counseled against being in places where an episode could cause serious injury to themselves or others. They should be advised against driving, swimming or operating heavy machinery [35]. Patients with hemodynamically significant VT should be advised against driving for at least 6 months from the last episode, even if an implantable cardioverter defibrillator (ICD) has been implanted [36].

In the event of syncope the best approach is to keep the head as low as possible while avoiding bending at the hip which may compromise the venous return. Patients should be counseled to assume the recumbent posture as soon as they recognize the premonitory symptoms. Patients are not allowed to stand up until the sensation of physical weakness has passed. Assuming erect posture too soon can result in a recurrent episode of syncope.

Increasing fluid and salt intake is the first line treatment of patients with neurocardiogenic syncope. Increased fluid intake (2 liters in morning and more to keep urine clear during the day) was reported to reduce the frequency of episodes and increased sodium intake (7g /day) was found to increase blood pressure and plasma volume. [37, 38] If the presyncopal symptoms persist on the above intervention, the next line is to try isometric exercises of the arm and leg muscles. [39, 40] Tilt Training has been mostly evaluated in adolescent patients with neurocardiogenic syncope. These studies were done on a structured in-hospital and home tilt training program. [41, 42] At home the usual training is standing against the wall for 30-40 minutes once or twice a day. Although found to be quite effective, the long term compliance for this method appears to be poor [43].

Pharmacological Therapy

Various pharmacotherapeutic agents that are useful in the treatment of syncope are listed in table 5. Beta adrenergic blockers have been traditionally used for the treatment of neurocardiogenic syncope. The mechanism of action is believed to be due to reduced activation of left ventricular mechanoreceptors and blunting of the response to the increased adrenergic stimulation that occurs just prior to syncope. Randomized placebo-controlled trials of beta-blockers have yielded conflicting results. [44-50] The multicenter Prevention of Syncope Trial (POST) was a randomized, placebo-controlled, double-blind, trial designed to assess the effects of metoprolol in vasovagal syncope over a 1-year treatment period. [50] Of 208 patients studied in this trial, 108 were randomized to metoprolol arm (at highest-tolerated doses from 25 to 200 mg daily), and 100 to placebo arm. The primary outcome - likelihood of recurrent syncope was not significantly different between the 2 groups. In a prespecified subgroup of patients older than 42 years, a nonsignificant trend towards benefit was observed (hazard ratio: 0.61; 95% CI 0.35 to 1.09, P=0.076), but was no better than placebo in those younger than 42 years of age. The reason for this difference is unknown.

Midodrine, a direct alpha-1 adrenergic receptor agonist approved in the United States for the treatment of orthostatic hypotension is also been used for the treatment of neurocardiogenic syncope. Midodrine has been found to have beneficial effects on symptoms, tilt table test and quality of life. [51, 52] However, midodrine can also worsen supine hypertension in these patients.

Fludrocortisone is a synthetic mineralocorticoid which promotes plasma volume expansion and also causes vasoconstriction of peripheral vasculature. It has been reported to be effective in a randomized trial of pediatric patients with vasovagal syncope. Currently there are no randomized studies evaluating the usefulness of fludrocortisone in adult patients with syncope. The ongoing Second Prevention of Syncope Trial (POST II) is investigating the benefits of fludrocortisone in this population. Fludrocortisone can be used in frequently symptomatic patients in the absence of contraindication.

Table 5. Pharmacotherapy of reflex syncope

Beta blocker	Decreased cardiodepressor response, Blocks serotonin receptor Proven to be beneficial in age > 42 years
Fludrocortisone	Increased intravascular volume
Paroxetine	Serotonin reuptake inhibitor
Dextroamphetamine, Methylphenidate, Midodrine	Alpha stimulation
Bupropion	Norepinephrine and dopamine reuptake inhibitor
Clonidine	In autonomic failure patients
Pyridostigmine	In postural orthostatic hypotension syndrome
Octreotide	Splanchnic vasoconstriction
Scopolamine	Vagolytic
Disopyramide	Vagolytic
Carbamazepine	In glossopharyngeal neuralgia
Theophylline	Central and peripheral adrenergic agonist

Serotonin specific reuptake inhibitors are generally not indicated for routine use in patients with syncope. [53, 54] They are thought to mitigate the serotoninergic actions in the brain that could contribute to vasovagal syncope. In a randomized placebo controlled trial 82% of patients receiving paroxetine were syncope free at 25 months compared to 53% in the placebo group [53].

Recent studies have shown that pyridostigmine bromide, a cholinesterase inhibitor, significantly improves standing blood pressure in patients with orthostatic hypotension without worsening supine hypertension. [55, 56] Pyridostigmine may also help patients with POTS. [57] Other therapies have been suggested in patients with neurocardiogenic syncope. Agents like disopyramide, enalapril, theophylline and ephedrine need to be evaluated in controlled trials. Transdermal scopolamine was not found to be superior to placebo in a randomized control trial. [58] In uncontrolled studies in patients with syncope due to psychological stressors, biofeedback has been effective in desensitizing patients and to prevent recurrence [59].

Device Therapy

The indications for pacemakers and ICDs in a patient with syncope are presented in table 6. Symptomatic bradycardia resulting from sinoatrial pauses or advanced second degree or third degree heart block usually requires the implantation of a permanent pacemaker. However, reversible causes like drugs, toxins and infections like lyme disease that cause or contribute to the bradyarrhythmia may improve once the offending agent is withdrawn or treated appropriately. The role of permanent cardiac pacing in vasovagal syncope has been controversial [60].

Cardiac Pacing has been proposed as a treatment option for patients with vasovagal syncope based on the observation that one third of the patients have symptomatic bradycardia or asystole during tilt table testing or during spontaneously occurring syncope. Early studies of pacemaker implantation in vasovagal syncope showed promising results [61-63]. However, the Vasovagal Pacemaker Study II (VPS II) trials and vasovagal syncope and pacing (SYNPACE) trials, which compared the effects of pacemakers turned "on" or "off," found little overall difference in the time to first syncopal recurrence between the two groups. [64, 65] In the SYNPACE trial, however, the group that had asystole during preimplant tilt table testing showed a significant increase in time to first syncopal recurrence, as opposed to those who only experienced bradycardia. [65] These studies have suggested that permanent pacing should not be used as first-line therapy in vasovagal syncope, but rather it should be considered in those not responsive to or intolerant of other forms of therapy. In vasovagal syncope, the fall in blood pressure usually occurs before the fall in heart rate. This may explain the ineffectiveness of pacing based on rate criteria alone in patients with vasovagal syncope. New sensor technologies that allow for blood pressure measurement may change this picture in the future. The recent ISSUE II trial reported that permanent pacing in patients with periods of asystole had a significant reduction in the frequency of syncope. [66] Patients with the cardioinhibitory or mixed forms of carotid sinus hypersensitivity may benefit also from permanent pacing. In SAFE PACE, permanent pacing reduced falls, recurrent syncope, and injuries in elderly patients with frequent nonaccidental falls and cardioinhibitory carotid sinus hypersensitivity. [67] "Pacemaker syndrome" is rectified by changing from ventricular pacing to AV sequential pacing when hypotension is a result of loss of contribution of atrial systole to ventricular filling.

Pacemaker-mediated tachycardia can usually be corrected by reprogramming the pacemaker. ICDs are the first-line therapy for ventricular tachycardia in the setting of structural heart disease. Catheter ablation may be effective for right ventricular outflow tract tachycardia, bundle-branch reentrant tachycardia, fascicular tachycardias, and idiopathic left ventricular tachycardia. Catheter ablation is also a successful mode of therapy in patients with symptomatic AV nodal reentry tachycardia, atrial flutter, and some atrial tachycardias.

REFERENCES

[1] Kapoor WN. Current evaluation and management of syncope, *Circulation* 2002;106:1606-1609.

[2] Quinn J, McDermott D, Kramer N, Yeh C, Kohn MA, Stiell I, Wells G. Death after emergency department visits for syncope: how common and can it be predicted? *Ann. Emerg. Med.* 2008;51:585-590.

[3] Sun BC, Emond JA, Camargo CA,Jr. Direct medical costs of syncope-related hospitalizations in the United States, *Am. J. Cardiol.* 2005;95:668-671.

[4] Soteriades ES, Evans JC, Larson MG, Chen MH, Chen L, Benjamin EJ, Levy D. Incidence and prognosis of syncope, *N. Engl. J. Med.* 2002;347:878-885.

[5] Somers VK, Abboud FM. Neurocardiogenic syncope, *Adv. Intern. Med.* 1996;41:399-435.

[6] Kerr SR, Pearce MS, Brayne C, Davis RJ, Kenny RA. Carotid sinus hypersensitivity in asymptomatic older persons: implications for diagnosis of syncope and falls, *Arch. Intern. Med.* 2006;166:515-520.

[7] Kaufmann H. Consensus statement on the definition of orthostatic hypotension, pure autonomic failure and multiple system atrophy, *Clin. Auton. Res.* 1996;6:125-126.

[8] Grubb BP, Kanjwal Y, Kosinski DJ. The postural tachycardia syndrome: a concise guide to diagnosis and management, *J. Cardiovasc. Electrophysiol.* 2006;17:108-112.

[9] Aronow WS, Ahn C. Postprandial hypotension in 499 elderly persons in a long-term health care facility, *J. Am. Geriatr. Soc.* 1994;42:930-932.

[10] Aronow WS, Ahn C. Association of postprandial hypotension with incidence of falls, syncope, coronary events, stroke, and total mortality at 29-month follow-up in 499 older nursing home residents, *J. Am. Geriatr. Soc.* 1997;45:1051-1053.

[11] Kessler C, Tristano JM, De Lorenzo R. The emergency department approach to syncope: evidence-based guidelines and prediction rules, *Emerg. Med. Clin. North Am.* 2010;28:487-500.

[12] Colivicchi F, Ammirati F, Melina D, Guido V, Imperoli G, Santini M, OESIL (Osservatorio Epidemiologico sulla Sincope nel Lazio) Study Investigators. Development and prospective validation of a risk stratification system for patients with syncope in the emergency department: the OESIL risk score, *Eur. Heart J.* 2003;24:811-819.

[13] Quinn JV, Stiell IG, McDermott DA, Sellers KL, Kohn MA, Wells GA. Derivation of the San Francisco Syncope Rule to predict patients with short-term serious outcomes, *Ann. Emerg. Med.* 2004;43:224-232.

[14] Reed MJ, Newby DE, Coull AJ, Jacques KG, Prescott RJ, Gray AJ. The Risk stratification Of Syncope in the Emergency department (ROSE) pilot study: a comparison of existing syncope guidelines, *Emerg. Med. J.* 2007;24:270-275.

[15] Serrano LA, Hess EP, Bellolio MF, Murad MH, Montori VM, Erwin PJ, Decker WW. Accuracy and quality of clinical decision rules for syncope in the emergency department: a systematic review and meta-analysis, *Ann. Emerg. Med.* 2010;56:362-373.e1.

[16] Brignole M, Ungar A, Bartoletti A, Ponassi I, Lagi A, Mussi C, Ribani MA, Tava G, Disertori M, Quartieri F, Alboni P, Raviele A, Ammirati F, Scivales A, De Santo T, Evaluation of Guidelines in Syncope Study 2 (EGSYS-2) GROUP. Standardized-care pathway vs. usual management of syncope patients presenting as emergencies at general hospitals, *Europace* 2006;8:644-650.

[17] Del Rosso A, Ungar A, Maggi R, Giada F, Petix NR, De Santo T, Menozzi C, Brignole M. Clinical predictors of cardiac syncope at initial evaluation in patients referred urgently to a general hospital: the EGSYS score, *Heart* 2008;94:1620-1626.

[18] Ungar A, Del Rosso A, Giada F, Bartoletti A, Furlan R, Quartieri F, Lagi A, Morrione A, Mussi C, Lunati M, De Marchi G, De Santo T, Marchionni N, Brignole M, Evaluation of Guidelines in Syncope Study 2 Group. Early and late outcome of treated patients referred for syncope to emergency department: the EGSYS 2 follow-up study, *Eur. Heart J.* 2010;31:2021-2026.

[19] Sule S, Palaniswamy C, Aronow WS, Ahn C, Peterson SJ, Adapa S, Mudambi L. Etiology of syncope in patients hospitalized with syncope and predictors of mortality and rehospitalization for syncope at 27-month follow-up, *Clin. Cardiol.* 2011;34:35-38.

[20] Reed MJ, Newby DE, Coull AJ, Jacques KG, Prescott RJ, Gray AJ. Role of brain natriuretic peptide (BNP) in risk stratification of adult syncope, *Emerg. Med. J.* 2007;24:769-773.

[21] Pfister R, Diedrichs H, Larbig R, Erdmann E, Schneider CA. NT-pro-BNP for differential diagnosis in patients with syncope, *Int. J. Cardiol.* 2009;133:51-54.

[22] Ammirati F, Colaceci R, Cesario A, Strano S, Della Scala A, Colangelo I, De Santo T, Toscano E, Ricci R, Santini M. Management of syncope: clinical and economic impact of a Syncope Unit, *Europace* 2008;10:471-476.

[23] Linzer M, Prystowsky EN, Brunetti LL, Varia IM, German LD. Recurrent syncope of unknown origin diagnosed by ambulatory continuous loop ECG recording, *Am. Heart J.* 1988;116:1632-1634.

[24] Zimetbaum PJ, Kim KY, Josephson ME, Goldberger AL, Cohen DJ. Diagnostic yield and optimal duration of continuous-loop event monitoring for the diagnosis of palpitations. A cost-effectiveness analysis, *Ann. Intern. Med.* 1998;128:890-895.

[25] Paruchuri V, Adhaduk M, Garikipati NV, Steinberg JS, Mittal S. Clinical utility of a novel wireless implantable loop recorder in the evaluation of patients with unexplained syncope, *Heart Rhythm.* 2011.

[26] Edvardsson N, Frykman V, van Mechelen R, Mitro P, Mohii-Oskarsson A, Pasquie JL, Ramanna H, Schwertfeger F, Ventura R, Voulgaraki D, Garutti C, Stolt P, Linker NJ, PICTURE Study Investigators. Use of an implantable loop recorder to increase the diagnostic yield in unexplained syncope: results from the PICTURE registry, *Europace* 2011;13:262-269.

[27] Sarasin FP, Louis-Simonet M, Carballo D, Slama S, Rajeswaran A, Metzger JT, Lovis C, Unger PF, Junod AF. Prospective evaluation of patients with syncope: a population-based study, *Am. J. Med.* 2001;111:177-184.

[28] Bonney ME, Reed MJ, Gray AJ. The Prediction of Risk In Syncope using ECG characteristics (PRISE) pilot study: can heart rate variability be used to predict risk in patients presenting to the emergency department with syncope? *Emerg. Med. J.* 2009;26:32-36.

[29] Zareba W, Moss AJ, le Cessie S, Hall WJ. T wave alternans in idiopathic long QT syndrome, *J. Am. Coll Cardiol.* 1994;23:1541-1546.

[30] Benditt DG, Ferguson DW, Grubb BP, Kapoor WN, Kugler J, Lerman BB, Maloney JD, Raviele A, Ross B, Sutton R, Wolk MJ, Wood DL. Tilt table testing for assessing syncope. American College of Cardiology, *J. Am. Coll Cardiol.* 1996;28:263-275.

[31] Grubb BP, Kosinski D. Tilt table testing: concepts and limitations, *Pacing Clin. Electrophysiol.* 1997;20:781-787.

[32] Strickberger SA, Benson DW, Biaggioni I, Callans DJ, Cohen MI, Ellenbogen KA, Epstein AE, Friedman P, Goldberger J, Heidenreich PA, Klein GJ, Knight BP, Morillo CA, Myerburg RJ, Sila CA, American Heart Association Councils on Clinical Cardiology, Cardiovascular Nursing, Cardiovascular Disease in the Young, and Stroke, Quality of Care and Outcomes Research Interdisciplinary Working Group, American College of Cardiology Foundation, Heart Rhythm Society, American Autonomic Society. AHA/ACCF Scientific Statement on the evaluation of syncope: from the American Heart Association Councils on Clinical Cardiology, Cardiovascular Nursing, Cardiovascular Disease in the Young, and Stroke, and the Quality of Care and Outcomes Research Interdisciplinary Working Group; and the American College of Cardiology Foundation: in collaboration with the Heart Rhythm Society: endorsed by the American Autonomic Society, *Circulation* 2006;113:316-327.

[33] Pietrucha AZ, Wnuk M, Wojewodka-Zak E, Wegrzynowska M, Mroczek-Czernecka D, Bzukala I, Konduracka E, Piwowarska W. Evaluation of sinus and atrioventricular nodes function in patients with vasovagal syncope, *Pacing. Clin. Electrophysiol.* 2009;32 Suppl 1:S158-62.

[34] Fujimura O, Yee R, Klein GJ, Sharma AD, Boahene KA. The diagnostic sensitivity of electrophysiologic testing in patients with syncope caused by transient bradycardia, *N. Engl. J. Med.* 1989;321:1703-1707.

[35] Sorajja D, Shen WK. Driving guidelines and restrictions in patients with a history of cardiac arrhythmias, syncope,or implantable devices, *Curr Treat Options Cardiovasc. Med.* 2010;12:443-456.

[36] Epstein AE, Baessler CA, Curtis AB, Estes NA,3rd, Gersh BJ, Grubb B, Mitchell LB, American Heart Association, Heart Rhythm Society. Addendum to "Personal and public safety issues related to arrhythmias that may affect consciousness: implications for regulation and physician recommendations: a medical/scientific statement from the American Heart Association and the North American Society of Pacing and Electrophysiology": public safety issues in patients with implantable defibrillators: a scientific statement from the American Heart Association and the Heart Rhythm Society, *Circulation* 2007;115:1170-1176.

[37] Younoszai AK, Franklin WH, Chan DP, Cassidy SC, Allen HD. Oral fluid therapy. A promising treatment for vasodepressor syncope, *Arch. Pediatr. Adolesc. Med.* 1998;152:165-168.

[38] El-Sayed H, Hainsworth R. Salt supplement increases plasma volume and orthostatic tolerance in patients with unexplained syncope, *Heart* 1996;75:134-140.

[39] Brignole M, Croci F, Menozzi C, Solano A, Donateo P, Oddone D, Puggioni E, Lolli G. Isometric arm counter-pressure maneuvers to abort impending vasovagal syncope, *J. Am. Coll Cardiol.* 2002;40:2053-2059.

[40] Krediet CT, van Dijk N, Linzer M, van Lieshout JJ, Wieling W. Management of vasovagal syncope: controlling or aborting faints by leg crossing and muscle tensing, *Circulation* 2002;106:1684-1689.

[41] Di Girolamo E, Di Iorio C, Leonzio L, Sabatini P, Barsotti A. Usefulness of a tilt training program for the prevention of refractory neurocardiogenic syncope in adolescents: A controlled study, *Circulation* 1999;100:1798-1801.

[42] Abe H, Kondo S, Kohshi K, Nakashima Y. Usefulness of orthostatic self-training for the prevention of neurocardiogenic syncope, *Pacing Clin. Electrophysiol.* 2002;25:1454-1458.

[43] Foglia-Manzillo G, Giada F, Gaggioli G, Bartoletti A, Lolli G, Dinelli M, Del Rosso A, Santarone M, Raviele A, Brignole M. Efficacy of tilt training in the treatment of neurally mediated syncope. A randomized study, *Europace* 2004;6:199-204.

[44] Brignole M, Menozzi C, Gianfranchi L, Lolli G, Bottoni N, Oddone D. A controlled trial of acute and long-term medical therapy in tilt-induced neurally mediated syncope, *Am. J. Cardiol.* 1992;70:339-342.

[45] Sheldon R, Rose S, Flanagan P, Koshman ML, Killam S. Effect of beta blockers on the time to first syncope recurrence in patients after a positive isoproterenol tilt table test, *Am. J. Cardiol.* 1996;78:536-539.

[46] Di Girolamo E, Di Iorio C, Sabatini P, Leonzio L, Barsotti A. Evaluation of the effects of diverse therapeutic treatments versus no treatment of patients with neurocardiogenic syncope, *Cardiologia* 1998;43:833-837.

[47] Madrid AH, Ortega J, Rebollo JG, Manzano JG, Segovia JG, Sanchez A, Pena G, Moro C. Lack of efficacy of atenolol for the prevention of neurally mediated syncope in a highly symptomatic population: a prospective, double-blind, randomized and placebo-controlled study, *J. Am. Coll Cardiol.* 2001;37:554-559.

[48] Ventura R, Maas R, Zeidler D, Schoder V, Nienaber CA, Schuchert A, Meinertz T. A randomized and controlled pilot trial of beta-blockers for the treatment of recurrent syncope in patients with a positive or negative response to head-up tilt test, *Pacing Clin. Electrophysiol.* 2002;25:816-821.

[49] Flevari P, Livanis EG, Theodorakis GN, Zarvalis E, Mesiskli T, Kremastinos DT. Vasovagal syncope: a prospective, randomized, crossover evaluation of the effect of propranolol, nadolol and placebo on syncope recurrence and patients' well-being, *J. Am. Coll Cardiol.* 2002;40:499-504.

[50] Sheldon R, Connolly S, Rose S, Klingenheben T, Krahn A, Morillo C, Talajic M, Ku T, Fouad-Tarazi F, Ritchie D, Koshman ML, POST Investigators. Prevention of Syncope Trial (POST): a randomized, placebo-controlled study of metoprolol in the prevention of vasovagal syncope, *Circulation* 2006;113:1164-1170.

[51] Perez-Lugones A, Schweikert R, Pavia S, Sra J, Akhtar M, Jaeger F, Tomassoni GF, Saliba W, Leonelli FM, Bash D, Beheiry S, Shewchik J, Tchou PJ, Natale A. Usefulness of midodrine in patients with severely symptomatic neurocardiogenic syncope: a randomized control study, *J. Cardiovasc. Electrophysiol.* 2001;12:935-938.

[52] Kaufmann H, Saadia D, Voustianiouk A. Midodrine in neurally mediated syncope: a double-blind, randomized, crossover study, *Ann. Neurol.* 2002;52:342-345.

[53] Di Girolamo E, Di Iorio C, Sabatini P, Leonzio L, Barbone C, Barsotti A. Effects of paroxetine hydrochloride, a selective serotonin reuptake inhibitor, on refractory vasovagal syncope: a randomized, double-blind, placebo-controlled study, *J. Am. Coll Cardiol.* 1999;33:1227-1230.

[54] Grubb BP, Wolfe DA, Samoil D, Temesy-Armos P, Hahn H, Elliott L. Usefulness of fluoxetine hydrochloride for prevention of resistant upright tilt induced syncope, *Pacing Clin. Electrophysiol.* 1993;16:458-464.

[55] Singer W, Opfer-Gehrking TL, McPhee BR, Hilz MJ, Bharucha AE, Low PA. Acetylcholinesterase inhibition: a novel approach in the treatment of neurogenic orthostatic hypotension, *J. Neurol. Neurosurg. Psychiatry* 2003;74:1294-1298.

[56] Singer W, Sandroni P, Opfer-Gehrking TL, Suarez GA, Klein CM, Hines S, O'Brien PC, Slezak J, Low PA. Pyridostigmine treatment trial in neurogenic orthostatic hypotension, *Arch. Neurol.* 2006;63:513-518.

[57] Raj SR, Black BK, Biaggioni I, Harris PA, Robertson D. Acetylcholinesterase inhibition improves tachycardia in postural tachycardia syndrome, *Circulation* 2005;111:2734-2740.

[58] Lee TM, Su SF, Chen MF, Liau CS, Lee YT. Usefulness of transdermal scopolamine for vasovagal syncope, *Am. J. Cardiol.* 1996;78:480-482.

[59] McGrady AV, Bush EG, Grubb BP. Outcome of biofeedback-assisted relaxation for neurocardiogenic syncope and headache: a clinical replication series, *Appl. Psychophysiol. Biofeedback* 1997;22:63-72.

[60] Sutton R. Has cardiac pacing a role in vasovagal syncope? *J. Interv. Card Electrophysiol.* 2003;9:145-149.
[61] Connolly SJ, Sheldon R, Roberts RS, Gent M. The North American Vasovagal Pacemaker Study (VPS). A randomized trial of permanent cardiac pacing for the prevention of vasovagal syncope, *J. Am. Coll Cardiol.* 1999;33:16-20.
[62] Sutton R, Brignole M, Menozzi C, Raviele A, Alboni P, Giani P, Moya A. Dual-chamber pacing in the treatment of neurally mediated tilt-positive cardioinhibitory syncope : pacemaker versus no therapy: a multicenter randomized study. The Vasovagal Syncope International Study (VASIS) Investigators, *Circulation* 2000;102:294-299.
[63] Ammirati F, Colivicchi F, Santini M, Syncope Diagnosis and Treatment Study Investigators. Permanent cardiac pacing versus medical treatment for the prevention of recurrent vasovagal syncope: a multicenter, randomized, controlled trial, *Circulation* 2001;104:52-57.
[64] Connolly SJ, Sheldon R, Thorpe KE, Roberts RS, Ellenbogen KA, Wilkoff BL, Morillo C, Gent M, VPS II Investigators. Pacemaker therapy for prevention of syncope in patients with recurrent severe vasovagal syncope: Second Vasovagal Pacemaker Study (VPS II): a randomized trial, *JAMA* 2003;289:2224-2229.
[65] Raviele A, Giada F, Menozzi C, Speca G, Orazi S, Gasparini G, Sutton R, Brignole M, Vasovagal Syncope and Pacing Trial Investigators. A randomized, double-blind, placebo-controlled study of permanent cardiac pacing for the treatment of recurrent tilt-induced vasovagal syncope. The vasovagal syncope and pacing trial (SYNPACE), *Eur. Heart J.* 2004;25:1741-1748.
[66] Brignole M, Sutton R, Menozzi C, Garcia-Civera R, Moya A, Wieling W, Andresen D, Benditt DG, Vardas P, International Study on Syncope of Uncertain Etiology 2 (ISSUE 2) Group. Early application of an implantable loop recorder allows effective specific therapy in patients with recurrent suspected neurally mediated syncope, *Eur. Heart J.* 2006;27:1085-1092.
[67] Kenny RA, Richardson DA, Steen N, Bexton RS, Shaw FE, Bond J. Carotid sinus syndrome: a modifiable risk factor for nonaccidental falls in older adults (SAFE PACE), *J. Am. Coll Cardiol.* 2001;38:1491-1496.

In: Comas and Syncope: Causes, Prevention and Treatment ISBN 978-1-62100-603-9
Editors: Eduardo Silva and Gabriel Cruz.

Chapter 5

Psychosocial Sequelae and Recovery of Idiopathic Pediatric Acquired Brain Injury: A Longitudinal Case Study

Elizabeth B. Hamilton and Diomaris E. Jurecska
George Fox University
Graduate School of Clinical Psychology, Newberg, OR, US

Abstract

This chapter presents the remarkable trajectory of a young boy who experienced an idiopathic [1]acquired brain injury (ABI) of unknown etiology followed by 22 months of coma prior to his gradual road towards recovery. This case is of particular interest because of available premorbid intellectual and achievement assessment data due to a mild pre-existing diagnosed learning disability. As such, it has been possible to compare benchmarks of his ongoing progress to his pre-illness cognitive and educational profile on a yearly basis since 2002 until the present. Previous research indicates that predictions of educational outcome after pediatric brain injury are a challenge. In this case, evidence-based interventions (EBI) are shown to maximize the student's intellectual and psychosocial potential. Although this young man resides in a rural, underserved community, multisystemic interventions have been successfully implemented in order to allow him to continue successfully in his home school district. The provision of multiple psychoeducational services have resulted in steady academic growth, although significant impairment persists in some domains.

The main purpose of this chapter is to examine potential mechanisms underlying the complex relationship between the sequelae of pediatric ABI and intellectual functioning as illustrated by this case study. Our secondary goal is to increase awareness of the role of established EBI in the management of patients with ABI at both the individual and the community level.

---

[1] For the purposes of this chapter and consistent with the literature, the terms Acquired Brain Injury (ABI) and Traumatic Brain Injury (TBI) are used interchangeably.

INTRODUCTION

The clinical and behavioral sequelae of a pediatric acquired brain injury (ABI) are typically severe, necessitating multiple life readjustments. Common documented struggles that follow ABI include impaired functioning in cognitive realms, academic achievement, language use, attention regulation, executive functioning, and adaptive behavior (Semrud-Clikeman, 2001). Attention problems post-injury have been shown to affect multiple processing areas, including vigilance, sustained attention, selective attention, inhibition, and shifting (Wassenberg, Max, Lindgren, and Schatz, 2004; Yeates, et al., 2005). Poorer outcomes have been found to be associated with disinhibition, irritability, emotional dyscontrol, social skills deficits, and depression (Ylvisaker, et al., 2007). Severe pediatric TBI injuries are especially associated with marked increases in behavioral and mood disorders, less adaptive daily functioning, and worse educational trajectories, in comparison to their cohort with mild to moderate injuries (Anderson, et al., 2006).

ABI within the pediatric population is particularly devastating because it not only adversely impacts the child's developmental stage at the time of the injury, but also has cumulative negative implications for interruptions in the child's development, interfering with previously acquired skills, while also inhibiting acquisition of unfolding developmental tasks. Because TBI is one of the leading causes of acquired disability for children aged 5 to 14 (Anderson, Catroppa, Morse, Haritou, and Rosenfeld, 2005), it is urgent that timely and effective rehabilitation strategies are developed to maximize recovery, enabling the child to most adaptively continue on a positive developmental trajectory.

TRAUMATIC BRAIN INJURY: A REVIEW OF INCIDENCE, PREVALENCE, AND RISK FACTORS

Traumatic brain injury (TBI) "is defined as a change in brain function that manifest as a confusion, altered level of consciousness, coma, seizure, acute sensory or motor neurological deficit, neuropsychological deficit, or behavioral change" (Gatzke-Kopp and Shannon, 2008, p. 208). A brief summary of the literature portrays emerging themes on the incidence, prevalence, and risk factors regarding ABI. The Center for Disease Control (CDC) estimates that approximately 474,000 children (ages 0-14) visit an emergency department annually for TBI (2010). In the United States child and adolescent mortality is significantly correlated to TBI since it is the number one cause of death or long-term disability in that population (Babikian and Asarnow, 2009). Epidemiological data have consistently shown predictors of TBI occurrences in children and adolescents. It appears that males from lower socioeconomic status between school and teenage years are twice as likely to acquire a TBI when compare to their female counterparts (Gatzke-Kopp and Shannon, 2008; Patrick, Patrick, and Duncan, 2006).

Overall, a linear correlation between the severity of injury and long-term disability and mortality suggests that injuries on the severe end of the spectrum are significantly correlated with higher levels of impairment and mortality (Kirkwood et al., 2008; Yeates et al. 2001). Generally, in the literature the acute injury severity is measured by the Glasgow Coma Scale (GCS; Jennett and Teasdale, 1981), which consistently demonstrates valid concurrent validity properties (Chia-Ying et al., 2006). The score ranges are defined as follows: mild (13-15 GCS), moderate (9-12 GCS), and severe (3-8 GCS; Babikian and Asarnow, 2009).

Relevant Case Studies

The transition and recovery of TBI has its origin in the legendary case of Phineas Gage. In 1848, a railroad worker survived an accident that permanently changed his personality (see Daffner and Searl, 2008). Unlike other TBI patients, Gage's fame resided in the uniqueness of his case. Similar to Gage's story, the current case of Daniel is one of a kind. An extensive review of the literature did not reveal a similar case study because of its exceptional characteristics and multicausality: a combination of seizure, high fever, and prolonged coma.

Present Case Synopsis

Daniel was born in 1995; he is the third of four surviving children; he has two older sisters and a younger brother. Daniel's parents each has a high school level education. His mother works as a full time homemaker, and his father fabricates farm equipment. Daniel's parents are married and he has always lived with them. Per mother's report, Daniel's early years were "normal"; he reached developmental milestones appropriately and did not present with any atypical illness. Daniel and his mother both reported that his interests pre and post trauma are related to cars and mechanical tasks with motors and engines. In October of 2004, at age 9, Daniel "fell to the floor in his gym class, became ill, and had a seizure." According to medical records, within five days Daniel was admitted to a local children's hospital . Within a short period of time, he lapsed into coma becoming non-responsive. Daniel was functioning at a GCS-III, and all medical providers agreed that he was in a "persistent vegetative state." At this point medical records indicate that Daniel developed encephalitis of unknown etiology. At the end of a month, he was transferred to a pediatric care facility, functioning at a Rancho Los Amigos level of III-Localized Response (Rancho Los Amigos National Rehabilitation Center, 2011). During his stay at this facility his seizure activity stopped, but he remained unable to communicate, move, or respond to one- on- one commands. In March 2006, he was noted to respond to noise by showing a change in breathing patterns and agitation. In September of 2006, he was awake and started to demonstrate voluntary motor activity. His feeding tube was removed, and he was placed at a local children's hospital for inpatient rehabilitation for a 2-month period. In December 2006 he returned home.

A review of Daniel's school records prior to the coma revealed that in 2000 Daniel was found eligible for communication services based on articulation errors. Daniel was dismissed from speech sessions in May, 2002 because he no longer met criteria for services. He was later assessed for a learning disability and met eligibility criteria for a reading disorder while in second grade. Daniel's case study is particularly pivotal, because premorbid intellectual and academic data are available due to the child's previously documented reading disorder. Daniel's current Individualized Education Plan (IEP) is flexible, considered to be a "working document" both by the parents, and the teachers. He has a note taker, and a special aide to work on his assignments. Other accommodations include: repeating instructions and directions, having consistent rules and routines, keeping distractions at minimum, providing support in organizational strategies to reinforce staying on task, and accessing a behavioral plan on a need basis. In an earlier article, we compared in detail this young man's cognitive profile and academic achievement pattern prior and subsequent to the child's TBI (Jurecska, Hamilton, and McConnell, 2011). The present chapter summarizes the boy's continued developmental trajectory in multiple domains as he transitions into high school, with ongoing assessment of both cognitive and psychoemotional factors. In addition, an integrative model is presented in which multisystemic integrative factors have contributed to the patient's ongoing rehabilitation and reintegration into his home community.

Common Sequelae of Acquired Brain Injury in the Pediatric Population

There is wide heterogeneity in the studied population of children with ABI, including differences in injury severity, age of trauma, and remediation programs, which contribute to variability in outcome studies. The importance of delineating between multiple etiologies and severity level when predicting post-morbid outcome is highlighted in a recent review by Johnson, DeMatt, and Salorio (2009). For the purposes of the current chapter, the authors concentrate primarily on research conducted on children with more severe injury, as treatment implications which are effective in this group may be assumed to also have efficacy in more moderate TBI cases.

Cognitive Adaptation Post-ABI

Across severity groups, neurocognitive domains most affected by TBI have been found to be intellectual functioning, processing speed, attention, and verbal memory (Babikian and Asarnow, 2009). Within IQ measures, visual-perceptual skills tend to emerge as more resilient than verbal comprehension skills. The unfolding consequences of injury are seen as deficits in problem solving and cognitive fluency at follow-up, while domains showing the most recovery include intellectual skills, primarily within perceptual domains, and processing speed.

Within the most severe injury group, marked residual symptoms were found at initial evaluation, less than six months following injury, including very large effects for intellectual functioning, executive functioning, and verbal immediate and delayed memory. Adverse commulative effects were seen by the time of two year follow-up, as impairments in all assessed neurocognitive domains were found (Babikian and Asarnow, 2009).

Intellectual Functioning

The Wechsler Intelligence Scale for Children, 4th edition (WISC-IV; Wechsler, 2003) was used to assess Daniel's intelligence in 2009, both because it is highly reliable and because this is congruent with the IQ test that Daniel was given during his assessment for a reading disorder in 2002. A comparison of Daniel's premorbid abilities and post-coma functioning as assessed by the WISC indices is provided in Table 1.

Table 1. Comparison of WISC-III and WISC-IV Across Time

Composite	WISC-III (12/02, age 7)			WISC-IV (4/09, age 13)		
	IQ/Index Scores	Percentiles	Confidence Intervals (95%)	IQ/Index Scores	Perc-entiles	Confidence Intervals (95%)
VCI	101	53rd		71	3rd	66-80
PRI	96	40th		88	21st	81-97
WMI	not part of test	--	--	59	0.3	55-70
PSI	not part of test	--	--	56	0.2	52-70
IQ	99	47th		63	1st	59-69

Note. VCI = Verbal Comprehension Index, PRI = Perceptual Reasoning, WMI = Working Memory Index, PSI = Processing Speed Index, and IQ = Full Scale IQ. Adapted from Educational implications following idiopathic encephalopathy and prolonged coma: A longitudinal case study (2011). *Journal of Research In Special Education Needs* by: Jurecska, Hamilton, and McConnell, C.

Table 2. WISC-IV Subtests

	12/02	4/09		12/02	4/09
Block Design	11	7	Matrix Reasoning	N/A	4
Similarities	12	6	Comprehension	12	6
Digit Span	N/A	5	Symbol Search	N/A	1
Picture Concepts	N/A	13	Information	8	5
Coding	5	3	Arithmetic	10	1
Vocabulary	9	3	Picture Arrangement	6	N/A
Letter-Number Sequencing	N/A	1	Picture Completion	13	N/A

Note. Adapted from Educational implications following idiopathic encephalopathy and prolonged coma: A longitudinal case study (2011). *Journal of Research In Special Education Needs* by: Jurecska, Hamilton, and McConnell, C.

In order to more fully elaborate this child's ongoing cognitive development, the Woodcock Johnson Tests of Cognitive Abilities, III (WJ-III; Woodcock, McGrew, and Mather, 2001b) was used as the intellectual assessment instrument for his most recent evaluation in 2011. The WJ III provides indices which are congruent with many of those utilized in the WISC-IV, while incorporating additional domains which may be linked in different ways to intellectual skills. See Table 2 for Daniel's WJ III profile.

Consistent with Babikian and Asarnow (2009), Daniel's non-verbal reasoning skills seem to be a relative strength compared to his verbal skills. However, there is noteworthy variability within the non-verbal reasoning domain, with some tasks being performed age appropriately, others showing significant impairment, and still others falling between these extremes. Daniel's performance in WISC-IV domains of memory and processing speed will be noted below. Compared to the administration of a WISC-III in December, 2002, it would seem, compared to his peers, Daniel is functioning at a significantly lower level. While it is not possible to directly compare results from the WISC-IV and the WJ Cognitive-III, trends in similar areas of cognitive development may be noted. Daniel's overall low intellectual functioning is congruent with that observed 5 years postinjury in children with severe TBI (Catroppa, et al., 2008). Although results from the most recent intelligence assessment reveal overall intellectual ability, as indexed by the WJ III GIA, in the low range of standard scores, Daniel currently evidences age-normative skills in the domains of both comprehension knowledge and visual-spatial thinking.

Table 3. Woodcock-Johnson III Cognitive Profile (6/11)

WJ Cognitive Cluster	Standard Score	Classification
Comprehension Knowledge	90	Average
Long-term Retrieval	85	Low average
Fluid Reasoning	51	Very low
Processing Speed	60	Very low
Short-term Memory	61	Very low
Visual-spatial Thinking	96	Average

Table 4. Wide Range Assessment of Memory and Learning, 2nd Edition (WRAML-2)

WRAML-2	Scale Score Date: (10/07) Age: 12	Scale Score Date: (04/09) Age: 13
Story Memory	7	8
Story Memory Delayed	8	7
Verbal Learning	1	6
Number-Letter	3	5
Finger Windows	1	5
Verbal Learning Recall	12	4
Design Memory	9	6

Academic Skills

When assessed in the fifth grade, Daniel's skills in the mechanics of reading (decoding), spelling and arithmetic range from first to mid third grade. With respect to spelling, Daniel showed difficulties especially with non- phonetic words. In math, Daniel demonstrated inconsistent mastery of single and double digit addition and subtraction problems. Occasionally, he would neglect the plus and minus signs. No multiplication or division facts seemed to have been mastered.

Follow-up assessment in the eighth grade indicated small, gradual academic gains, with his overall scholastic achievement falling at mid-third grade level. There is currently little variability within his academic performance, with broad reading skills at the 3.3 grade level, broad math at the 3.0 grade level, and broad written language at the 3.3 grade level. It should be noted, however, that he has particular difficulty with educational tasks demanding fluency, as his reading fluency falls at 2.6, math fluency at 2.7, and written fluency at 1.8 grade equivalent. These relative deficits are congruent with his very slow processing speed.

Memory and Learning

Generally, Daniel's verbal and non- verbal immediate memory skills were found within the borderline range compared to age-mates. With meaningful verbal information, Daniel seemed able to retain over time what he has learned; however, with less meaningful verbal information (such as learning isolated terms), his learning rate was marginal and he seemed to forget more information than he did with meaningful material. Compared to the administration of a **Wide Range Assessment of Memory and Learning, Second Edition (WRAML-2**; Sheslow, and Adams, 2003) completed 20/2007, Daniel generally showed a similar level of verbal and visual skills although more consistency across strengths and weaknesses. This child's continued struggles with working memory, even five years after his recovery of consciousness, are congruent with the persistence of memory deficits documented in children with severe TBI, in contrast to the resolution of memory issues at follow-up for youth with more moderate trauma (Roncadin, et al., 2004).

Psychological, Social and Behavioral Adaptation Following ABI

Post-injury difficulties within the psychosocial realm include irritability, aggression, social deficits, and depressive symptoms (Ylvisaker, et al., 2007). The importance of proactively addressing these symptoms is shown by studies indicating that these variables interfere with longitudinal post-injury quality of life even more than do neurocognitive deficits (Nybo, Sainio, and Muller, 2004). Children with TBI have been found to have lower self-esteem and higher rates of maladaptive behavior, while describing themselves as less socially competent than controls (Andrews, Rose, and Johnson, 1998. In particular, children with severe brain injury have been found to have marked deficits in social problem solving (Janusz, Kirkwood, Yeates, and Taylor. 2002). Finally, the incidence of newly emerging behavioral disorders following pediatric ABI has been estimated to range between 35% and 70% (Ylvisaker, et al., 2007).

In 2008, Daniel's teacher reported that although he did not demonstrate inadequate behaviors or emotional dysregualtion in the classroom, he struggled with both inattentiveness and hyperactivity. By Daniel's self-report, he described himself at risk for both anxiety and issues with inadequacy. Follow-up data was collected on the Behavior Assessment System for Children, second edition, (BASC-2) (Reynolds and Kamphaus, 2004) in the late spring of 2011, at which time this young man was turning sixteen years of age. The BASC-II was completed by his special education teacher, who has had extensive ongoing contact with Daniel over the past five years. His teacher reported no externalizing problems across domains, and the only internalizing problem area on which she rated him as slightly elevated, falling in the "at-risk" classification (T = 60), was for somatization concerns. In contrast, within the domain of school-based problems, "clinically significant" difficulties were reported for learning problems (T = 79), with difficulty comprehending and completing schoolwork in a variety of subject areas, with "at-risk" difficulties for attentional problems (T = 61). Congruent with his post-morbid cognitive profile, Daniel's teacher-reported BASC-II placed him in the "at-risk" classification for overall adaptive skills. However, it should be noted that within this overarching domain, he is viewed as having good social skills, adequate study skills, and the ability to work under pressure. The adaptive skills components which are problematic for him include adaptability to changes (T = 37, "at-risk") and functional communication (T = 27, "clinically significant"), with difficulties reported both for expressive and receptive communication skills. Daniel's self-report of behavioral functioning was modified due to his cognitive profile, and as such he completed the self-report version of the BASC-2 for ages 8 – 11. His responses generally portray a fairly well-adapted young man, who endorses items reflecting a positive attitude towards school, sociability, and a positive sense of self. His completion of the short form of the Child Depression Inventory (Kovacs, 1992) showed that he did not endorse any of the items designed to capture a range of depressive symptoms. The Vineland Adaptive Behavior Scale (Sparrow, Balla, and Cicchetti, 1984) was used to assess Daniel's social and adaptive functioning across multiple domains, as observed by Daniel's mother. According to maternal report, Daniel experiences significant difficulty with emotional regulation and also has problems both initiating and monitoring his own performance. His adaptive behavior composite score was 65, which places him in the first percentile of overall function compared to his peers. This means that Daniel is in the low range of ability for being able to meet age standards for daily activities and tasks related to taking care of himself. Daniel also scored in the low range in communication, daily living, and socialization, which suggests that his development in these areas is delayed compared to his peers. A low level of adaptive skills is consistent with his premorbid diagnosis of mild mental retardation. Daniel's psychobehavioral profile is different from the typical trajectory for children with severe TBI, as research has found that clinically significant behavioral problems and social impairments are more likely to be present at 4 years postinjury in those with severe TBI as contrasted to those with mild or moderate injuries (Catroppa, et al., 2008; Schwartz, et al., 2003; Yeates et al., 2004). While not ignoring the negative effect of Daniel's ABI on his adaptive functioning, particularly in the realm of communication, to date he has not developed clinically significant behavioral or emotional issues. The authors propose that multiple protective factors as described below may have served to mediate the typically adverse impact of brain injury on this young man's behavioral adaptation.

Protective Factors Contributing to Positive Rehabilitative Outcome

It is clear that the most optimized post-ABI prognosis necessitates the maximization of multiple pre-existing factors as well as the maximization of available resources in the child's current environment. Some of the variables which have been found to mediate or potentiate recovery are discussed below, in particular relation to Daniel's case.

Age of Injury. Although cumulative research has yielded variable findings, increasing evidence indicates that earlier age at trauma is linked to more negative outcomes (Anderson, 2005; Campbell et al., 2004). Although the younger brain has been found to demonstrate more functional plasticity due to compensatory neural networks that may still be forming, brain injury prior to age 6 may be particularly devastating because the young child has yet to master fundamental skills linked to linguistic ability and problem solving. As such, the older child may be able to recover some previously acquired skill sets, which the younger child has yet to master, with subsequent acquisition being hindered by the complex neurocognitive post-injury sequelae. Data suggest that TBI after age 6 is particularly devastating as it is likely linked to interruption of fundamental skills acquisition, such as linguistic fluency and the attainment of operational processes. In a prospective longitudinal study of 122 children with TBI, Anderson (2005) found evidence for a "double-hazard" model for the trajectory of children sustaining severe injuries prior to age 3, supporting the early vulnerability theory.

Since Daniel's ABI occurred at the age of 9, he had already had the opportunity to master multiple developmental tasks, including self-care, language, reading, and social skills. As such, even after a very prolonged coma, a foundational skill level of adaptive functioning remained. In fact, it should be noted that part of Daniel's cognitive and academic decreases are likely related to difficulties with attaining new milestones and mastering progressive concepts, so that post-illness scores in some domains are similar to his pre-illness scores, but now are rated lower relative to peers. Research also has indicated cumulative unfolding of deficits, highlighting the need for effective interventions to be available throughout the lifespan. For example, as Daniel progresses beyond high school into early adulthood, intervention will need to be tailored to enhance independent living skills and fundamental job training.

High Levels of Family Cohesion. Strong family commitment and low levels of discord have been linked to more positive rehabilitative outcomes for children with ABI (Semrud-Clikeman, 2010). This is congruent with the wide body of research to date documenting the association between low cohesion and high conflict in the family environment with a wide range of child psychopathologies (George, Herman, and Ostrander, 2006). Further, within a large sample of children with intellectual disabilities, family dysfunction uniquely predicted the development of psychopathology (Wallander, Dekker, and Koot, 2006). Although Daniel is to date showing no diagnosable clinical syndromes, research indicates that TBI is often linked to difficulties with attention and dyscontrol, which may contribute to the onset or exacerbation of psychological and behavioral issues. As such, the warm and accepting home environment provided by his extended family may serve the purpose of both: (a) facilitating neurocognitive rehabilitation, and (b) buffering Daniel from the subsequent development of extensive behavioral problems.

"Optimistically realistic" expectations of family, school, and community partners. Emerging research suggests that the primary predictor for positive outcome may not be simply the severity of the injury itself, but rather the positive adaptability of the child's environment following ABI (Taylor, et al., 2002; Wade, Carey, and Wolfe, 2006). Because of the extensive amount of time the child spends within the educational system, it is crucial that the family and school develop a positive, collaborative working relationship in which appropriate school and community supports can be adjusted in an ongoing process of the child's academic and social trajectory (Semrud-Clikeman, 2010).

Daniel returned to his small, rural home school following recovery, and the positive working relationship among school professionals, parents and the child have resulted in ongoing adaptation of specialized services, contingent on his unfolding development. Because Daniel was previously on an individualized education plan (IEP), due to mild articulation errors difficulties and reading difficulties, he had already received ongoing psyhcoeducational evaluation, which has provided a premorbid index of his intellectual and achievement profile. This information has been invaluable in shaping ongoing school-based intervention. Examples of ways in which the educational system has specialized Daniel's placement include: (a) re-entry into classes a year below his pre-existing level in order to minimize deficits relative to peers, (b) the implementation of multiple learning resources; and (c) gradual entry into the high school system, with the selection of elective courses. The success of this school-based partnership is shown by the parents' adamant commitment to remaining at their home school in lieu of placing their son in a more specialized school, because "at his home school they understand him".

Injury Severity. Meta-analysis indicates that severity of injury is related not only to significant impairments across neurocognitive domains (Babikian and Asarnow, 2009), but also to increasing lags as the child progresses through subsequent developmental stages. The undetermined etiology of Daniel's unique case, with seizures and subsequent longlasting coma, place him in the most severe injury group represented in research studies, and as such, would predict poor prognosis with cumulative challenges. The fact that the home and school system have been able to provide an inclusive learning environment for Daniel and that he has been able to make a fairly positive psychological adjustment in spite of the very intensive nature of his ABI, further illustrates the potency of well-timed and comprehensive psychoeducational services.

INTERVENTION STRATEGIES FACILITATING COMMUNITY-BASED REHABILITATION

Although the bulk of the research on TBI intervention to date has focused primarily on the adult population, and as such may not be directly generalizable to pediatric rehabilitation, recent systematic research reviews suggest the efficacy of attentional training, long-term cognitive therapy, remediation of difficulties with visual-spatial processing, and the implementation of compensatory skills (Semrud-Clikeman, 2010;Slomine and Locascio, 2009). Ross, Dorris, and McMillan (2011) provide an excellent review article, which summarizes extant research on intervention efforts for children with acquired brain injury through March, 2010. In sum, some of the remediation strategies which have been shown to contribute to more adaptive outcomes include both: (a) direct intervention to retrain pre-existing neurocognitive functions, and (b) implementation of compensatory skill training, based on the utilization of intact neurological functions (Semrud-Clikeman, 2010).

Ylvisaker and colleagues (2007) provide a meta-analysis of behavioral interventions us in both child and adult studies, indicating the efficacy of both applied behavioral analysis (ABA) and positive behavior intervention focused on global lifestyle readjustments, frequently combined with training in executive functioning. A combination of cognitive behavioral interventions with positive behavior support has been found to improve the adaptation of two young children with severe TBI (Feeney and Ylvisaker, 2008). This converging data indicates that in order to maximize remediation programs, a multisystemic approach which utilizes all available intervention modalities with the child, family, school, and community should be implemented when possible.

A particularly salient factor unique to rehabilitative efforts in the pediatric population is reintegration into the school system. As research documents multiple difficulties in reintegrating the patient into his or her home community, attempts are made to maximize service delivery and reintegration (Johnson, DeMatt, and Salorio, 2009; Ylviasaker, et al., 2007). The current case is of particular interest because it illustrates the feasibility of optimal reintegration even within a rural community. The school system has collaboratively worked with the family to provide age-appropriate services for Daniel as he completes his education. The importance of utilizing positive behavioral interventions and ABA within a successful rehabilitative model is highlighted in a recent review article (Semrud-Clikeman, 2010).

In a recent study focused on attention remediation in a group of children and adolescents with TBI, Galbiati and colleagues (2009) found positive outcome effects for improvements in attention, but not overall intellectual functioning, as a result of specific remediation training, including a combination of a process-specific approach and metacognitive strategies. The pivotal role of attention in remediation is further highlighted by the finding that as attention skills improved, so did adaptive skills. Because the ability to attend to educational material and remain on-task is fundamental to the reintegration of the child with ABI into his or her school system, it has been important to equip Daniel with compensatory strategies for remaining attentive and enhancing concentration.

Finally, research has shown that the most effective rehabilitation efforts for TBI occur when multimodal interventions are embedded within a naturalistic environmental context (Ylvisaker, et al., 2005). Integrated coordination of care and external structured support has been available both in the home and school system for this young boy, thus maximizing the effect of management strategies and compensatory training.

Educational Intervention

Empirically Supported Classroom Interventions, Research indicates that there are a variety of modifications that can be made in the public school system that support the educational needs of students with ABI. Rees and Skidmore (2008) found that a classical classroom style enhanced learning opportunities for students with ABI. Their study of several individuals with ABI suggests that short lessons focused around one main theme with a predictable format and clear structure are especially conducive to the educational needs of these students. The repetition of a central idea throughout the lesson may be particularly helpful in compensating for the memory and processing issues faced by children with ABI. This is in contrast to the polyphonic classroom, in which a high level of attentiveness is needed to understand multiple layers of meaning surrounding a topic and to follow the relationships between a beginning idea and the conclusion of a lecture. Repetition of main themes, along with having a question and answer session at the end of lessons provided many opportunities for the students with ABI to hear and understand the main point of the lesson

(Rees and Skidmore, 2008). Clear transition points and organization are also helpful to students with ABI who may have impaired organizational skills.

Many of the modifications operationalized by Daniel's school system help remediate the difficulties he experiences with working memory, a common deficit of children with ABI from multiple etiologies (Slomine and Locascio, 2009). As Hanten and colleagues (2002) have documented, deficits in memory in children with ABI interfere with the ability to encode information in an efficient and organized manner, highlighting the importance of explicit instruction in organizational and mnemonic strategies. Finally, the centrality of attentional deficits for both the learning profile of this particular case study and children with ABI in general must be emphasized. The practice guidelines proposed by Laatsch and colleagues (2007) describe the documented efficacy of focused intervention within this domain.

Individualized Classroom Interventions for Daniel. Daniel's school has made a variety of accommodations to address his educational needs, in addition to the one-on-one aide and special education classes detailed in his educational history. These modifications in general education classes include tailoring curriculum, instruction, and expectations to Daniel's instructional level. For example, texts that are above his reading level are presented orally. In math, Daniel's instruction focuses on mastering mathematic operations, specifically learning to identify the correct operation within story problems. Teachers regularly review relevant vocabulary with Daniel prior to lessons, and he is given a written copy of class notes.

Global deficits in processing speed, typical of most children with ABI and learning disabilities (Calhoun and Mayes, 2005), are addressed with multiple modifications. For example, Daniel receives extra time to complete assignments, which are usually shortened, and he can take breaks from work if he becomes fatigued. To reduce motor skill fatigue, Daniel is encouraged to build his keyboarding and computer use skills. During class, he is allowed extra processing time to come up with oral answers. Tests and quizzes are modified to Daniel's instructional level, and he is allowed to answer orally or dictate answers. Speech assignments are also shortened and broken down. Daniel receives support in writing speeches out in sentences, so that he can focus on three main points, without using words that are difficult for him to pronounce. Finally, Daniel receives preferential seating near the teacher.

Developmental Trajectory for Continued Post-ABI Adjustment

Additional challenges have been documented for children with ABI as they transition to the middle and high school years, as the developmental tasks of autonomy and individuation are dependent on the ability to self-monitor, regulate impulses and emotions, and engage in long-term planning, all skills which are often adversely impacted by head trauma (Semrud-Clikeman, 2010).

Further, a review of the literature indicates that the "gap" between peers and students with ABI, particularly those with severe injuries, may widen as developmental requirements become more complex.

Some of the challenges faced in this young man's developmental trajectory include:

1. recent transition of his classmates into high school, while he himself has remained in 8th grade, attending an elective course at the high school, which conveniently is located next to the middle school complex
2. Daniel's younger brother has progressively made age typical developmental gains which have resulted in surpassing Daniel in several domains of functioning, a trend which will continue with the gap gradually widening

It is fortunate that this young man has multiple protective factors which research has shown to be positively associated with optimal post-ABI recovery. Although his injury was particularly severe, ongoing multisystemic intervention which builds on his cohesive family environment, positive psychoemotional characteristics, and post-injury retention of strong perceptual-reasoning skills may be expected to contribute to a relatively adaptive developmental trajectory. Follow-up of this intriguing case of pediatric ABI will be continued in collaboration with the family as this young man completes his high school years.

Conclusions

The current chapter details the relatively positive outcome of multifaceted, school-based intervention, even when the ABI has been severe, with initially very poor prognosis. As progressive research continues to evaluate optimal treatment protocols, based on the mediating factors of age at injury, severity of trauma, and compensatory factors, an ongoing focus should be on maximizing all intervention strategies within the child's home community. As the current case study illustrates, it is particularly important that the child be integrated as fully as possible within the school environment, with individualized educational adjustments effected in response to the child's developmental trajectory, especially in cases of more severe injury, due to the cumulative deficits found longitudinally. Although this chapter provides an intensive analysis of one child's ABI trajectory, it should be noted that his post-injury cognitive scores and adaptive scores are similar to those reported in larger scale studies (i.e., Galbiati, et al., 2009). While the authors' comments are limited due to the intrinsic nature of an individual case study, ongoing research entailing larger sample sizes should focus on combining the most effective evidence-based rehabilitation strategies with the most efficient and integrative means of service delivery for this specialized group of children.

Ethical Statement

Written informed consent for this article was obtained both from Daniel and from his parents. Initially the family explicitly requested that their son's actual name be used in order to give additional meaning to their family journey. However, because Daniel is still a minor, the researchers have used a pseudonym for future confidentiality issues. We appreciate the generosity and openness of Daniel and his family in allowing his unique experience to be shared in order to further educate those of us who work with children with special challenges.

AUTHOR NOTE

We warmly thank Daniel and his family for their willingness to share his unique case study. They are all heroes to us! In addition, we gratefully acknowledge the contributions of Heather MacKay and Courtney McConnell in assisting with the evaluation process.

REFERENCES

Anderson, V., Catroppa, C., Dudgeon, P., et al. (2006). Understanding predictors of functional recovery and outcome 30 months following early childhood head injury. *Neuropsychology*, *20*: 42-57.

Anderson, V., Catroppa, C., Morse, S., Haritou, F., and Rosenfeld, J. (2005). Attentional and processing skills following traumatic brain injury in early childhood. *Brain Injury, 19,* 699-710.

Anderson, V. (2005). Functional plasticity or vulnerability after early brain injury? *Pediatrics, 116 (6),* 1374-1382.

Andrews, , T. K., Rose, F. D., and Johnson, D. A. (1998). Social and behavioural effects of traumatic brain injury in children. *Brain Injury, 12,* 133-138.

Babikian, T., and Asarnow, R. (2009). Neurocognitive outcomes and recovery after pediatric TBI: Meta-analytic review of the literature. *Neuropsychology, 23 (3),* 282-296.

Calhoun, S. L. and Mayes, S. D. (2005). Processing speed in children with clinical disorders. *Psychology in the Schools*, *42 (4),* 333-343.

Campbell, C. G., Kuehn, S. M., Richards, P. M., Ventureyna, E., and Hutchinson, J. S. (2004). Medical and cognitive outcome in children with traumatic brain injury. *Canadian Journal of Neurological Science*, *32*: 213-219.

Catroppa, C., Anderson, V. A., Morse, S. A., et al. (2008). Outcome and predictors of functional recovery 5 years following pediatric traumatic brain injury (TBI). *Journal of Pediatric Psychology*, *7*, 707-718.

Center for Desease Control. (2010). Injury prevention and control: Ttraumatic brain injury. Retrieved from http://www.cdc.gov/TraumaticBrainInjury/ statistics.html

Chia-Ying, C., Chia-Ling, C., Pao-Tsai, C., Lai-Chu, S., Simon Fuk-Tan, T., and Wong, A. (2006). Critical score of Glasgow Coma Scale for pediatric traumatic brain injury. *Pediatric Neurology. 34*(5) 379-387, DOI: 10.1016/j.pediatrneurol.2005.10.012).

Daffner, K. R., and Searl, M. M. (2008). The dysexecutive syndromes. In M. Aminiff, F. Boller, and D. Swaab (Eds.). *Handbook of clinical neurology* (249-268). Amsterdam, The Netherlands: Elsevier.

Feeney, T. J., and Ylvisaker, M. (2008). Context-sensitive cognitive-behavioral supports for young children with TBI: A replication study. *Journal of Positive Behavioral Interventions, 10*, 115-128.

Galbiati, S., et al. (2009). Attention remediation following traumatic brain injury in childhood and adolescence. *Neuropsychology, 23(1),* 40-49.

Gatzke-Kopp, L. M., and Shannon, K. E. (2008) Brain injury as a risk for psychopathology. In T. P. Beauchaine and S. P. Hinshaw (Eds.) *Child and adolescent psychopathology* (208-233). Hoboken, NJ: Wiley.

George, C., Herman, K., and Ostrander, R. (2006). The family environment and developmental psychopathology: the unique and interactive effects of depression, attention, and conduct problems. *Child Psychiatry and Human Development*, *37*, 163-177.

Hanten, G., Zhang, L., and Levin, H. S. (2002). Selective learning in children after traumatic brain injury: a preliminary study. *Child Neuropsychology*, *8*, 107-120.

Ingoldsby, E. M., Shaw, D. S., and Garcia, M. M. (2001). Intrafamily conflict in relation to boys' adjustment at school. Development and Psychopathology, 13, 35-52.

Janusz, J. A., Kirkwood, M. W., Yeates, K. O., and Taylor, H. G. (2002). Social problem-solving skills in children with traumatic brain injury: Long-term outcomes and prediction of social competence. *Child Neuropsychology, 8*, 179-194.

Jennett, B., and Teasdale, G. (1981). *Management of head injuries*, 6th ed. Philadelphia: EA Davis.

Johnson, A. R., DeMatt, E., and Salorio, C. (2009). Predictors of outcome following acquired brain injury in children. *Developmental Disabilities Research Reviews*, *15*, 124-132.

Jurecska, D. E., Hamilton, E. B., and McConnell, C. E. (2011). Educational implications following idiopathic encephalopathy and prolonged coma: A longitudinal case study. *Journal of Research in Special Educational Needs, 11;* doi: 10.1111/j.1471-3802.2010.01180.x.

Kirkwood, M., Yeates, K. O., Taylor, H. G., Randolph, C., MacCrea, M., and Anderson, V. (2008). Management of pediatric mild traumatic brain injury: A neuropsychological review from injury to recovery. *Clinical Neuropsychology, 22*(5):769-800.

Kovacs, M. (1992). *Children's Depression Inventory*. Pearson Education.

Laatsch, L., Harrington, D., Hortz, G. et al. (2007). An evidence-based review of cognitive and behavioral rehabilitation treatment studies in children with acquired brain injury. *Journal of Head Trauma Rehabilitation*, *22*, 248-256.

Nybo, T., Sainio, M., and Muller, K. (2004). Stability of vocational outcome in adulthood after moderate-to-severe preschool brain injury. *Journal of the International Neuropsychological Society, 10,* 719-723.

Patrick, P. D., Patrick, S, T., and Duncan, E. (2006). Neuropsychological recovery: Children and adolescents following traumatic brain injury. In J. Leon-Carrion, K. R. Wild, and G. A. Zitnay (Eds.). Brain injury treatment: Theories and practices (studies on neuropsychology, neurology and cognition, (401- 440). New York, Taylor and Francis.

Rancho Los Amigos National Rehabilitation Center. (2011). *The Rancho Levels of Cognitive Functioning*. Retrieved from http://www.rancho.org/ Research_RanchoLevels.aspx.

Rees, S. A., and Skidmore, D. (2008). Redesigning the scaffolding metaphor to suit pupils with Acquired Brain Injury. European Journal of Special Needs Education, 23 (4), pp. 379-392.

Reynolds, C. R., and Kamphaus, R. W. (2004). *Behavioral assessment for children (2nd ed.).* Circle Pines, MN: American Guidance Service.

Roncadin, C., Guger, S., Archibald, J., et al. (2004). Working memory after mild, moderate, or severe childhood closed head injury. *Developmental Neuropsychology*, *25*, 21-36.

Ross, K., Dorris, L., and McMillan, T. (2011). A systematic review of psychological interventions to alleviate cognitive and psychosocial problems in children with acquired brain injury. *Developmental Medicine and Child Neurology*, online preview. Retrieved from Wiley Online.

Schwartz, H., Taylor, H. Drotar, D., Yeates, K. O., Wade, S., and Stancin, T. (2003). Long-term behavior problems following pediatric traumatic brain injury: Prevalence, predictors, and correlates *Journal of Pediatric. Psychology. 28*(4): 251-263 doi:10.1093/jpepsy/jsg013.

Sheslow, D., and Adams, W. (2003). Wide Range Assessment of Memory and Learning Second Edition administration and technical manual. Lutz, FL: Psychological Assessment Resources.

Semrud-Clikeman, M. (2001). *Traumatic brain injury in children and adolescents.* New York, NY: Guilford Press.

Semrud-Clikeman, M. (2010). Pediatric traumatic brain injury: Rehabilitation and transition to home and school. *Applied Neuropsychology, 17*, 116-122.

Slomine, B., and Locascio, G. (2009). Cognitive rehabilitation for children with acquired brain injury. *Developmental Disabilities Research Reviews, 15*, 133-143.

Sparrow, S., Balla, D., and Cicchetti, D. (1984). *Vineland Adaptive Behavioral Scales*. Circle Pines, MN: American Guidance Center.

Taylor, H.G., Yeates, K.O., Wade, S.L., Drotar, D., Stancin, T., and Minich, N. (2002). A prospective study of short- and long-term outcomes after traumatic brain injury in children: Behavior and academic achievement. *Neuropsychology*, *16*, 15-27.

Tomlin, P., Clarke, M., Robinson, G., et al. (2002). Rehabilitation in severe head injury in children: outcome and provision of care. *Developmental Medicine Child Neurology*, *44*, 828-837.

Wade, S. L., Carey, J., and Wolfe, C. R. (2006). An online family intervention to reduce parental distress following pediatric brain injury. *Journal of Consulting and Clinical Psychology, 74,* 445-454.

Wallender, J. L., Dekker, M. C., and Koot, H. M. (2006). Risk factors for psychopathology in children with intellectual disability: a prospective longitudinal population-based study. *Journal of Intellectual Disability Research*, *50* (4), 259-268.

Wassenberg, R., Max, J. E., Lindgren, S. D., and Schatz, A. (2004). Sustained attention in children and adolescents after traumatic brain injury: Relation to severity of injury, adaptive functioning, ADHD and social background. *Brain Injury, 18,* 751-764.

Wechsler, D. (2003). Wechsler Intelligence Scale for Children, 4th edition (WISC-IV). San Antonio, TX: Hardcourt Assessment.

Yeates, K. O., Armstrong, K., Janusz, J., Taylor, H. G., Wade, S., Stancin, T., and Drotar, D. (2005). Long-term attention problems in children with traumatic brain injury. *Journal of American Academy of Child and Adolescent Psychiatry, 44,* 574-584.

Woodcock, R. W., McGrew, K. S., and Mather, N. (2001b). *Woodcock-Johnson III Tests of Cognitive Abilities*. Itasca, IL: Riverside Publishing.

Yeates, K. O., Swift, E., Taylor, H. G., Wade, S. L., Drotar, D., Stancin, T., and Minich, N. (2004). Short- and long-term social outcomes following pediatric traumatic brain injury. *Journal of International Neuropsychology Society*, *10*, 412-426.

Yeates, K., Taylor, H., Barry, C., Drotar, D., Wade, S., Stancin, T. (2001). Neurobehavioral symptoms in childhood closed-head injuries: Changes in prevalence and correlates during the first year postinjury. *Journal of Pediatric Psychology. 26*, 79–91.

Ylvisaker, M., Turkstra, L., Coelho, C. A., Yorkston, K., Kennedy, M., Sohlberg, M. M., et al. (2007). Behavioral interventions for children and adults with behavior disorders after TBI: A systematic review of the evidence. *Brain Injury, 21,* 769-805.

Ylvisaker, M., Adelson, P. D., Braga, L. W., Burnett, S. M., Glang, A., Feeney, T. et al. (2005). Rehabilitation and ongoing support after pediatric TBI: twenty years of progress. *Journal of Head Trauma Rehabilitation*, *20*, 95-109.

In: Comas and Syncope: Causes, Prevention and Treatment ISBN 978-1-62100-603-9
Editors: Eduardo Silva and Gabriel Cruz. ©2012 Nova Science Publishers, Inc.

Chapter 6

MANAGEMENT OF NON TRAUMATIC COMAS IN SUB-SAHARAN AFRICA: CONTRIBUTION OF THE HEAD COMPUTERIZED TOMOGRAPHY SCAN

N. Kouamé*[*], A. M. N'goan-Domoua, A. N. Konan, A. Sétchéou, Z. O. D. Tra-Bi, R. D. N'gbesso and A. K. Kéita

Department of Radiology. Yopougon's Teaching Hospital. 21 BP 632 Abidjan 21

ABSTRACT

Objective: the aim of our study was to list the main etiologies of non traumatic comas in sub-Saharan Africa and to specify the importance of computed tomography scan in its management.

Patients and method: retrospective analysis of the files of 142 patients hospitalized for non traumatic coma, between January, 2006 and December, 2010, in the neurology, neurosurgery, emergency and intensive care departments of Yopougon Teaching Hospital (Abidjan-Ivory Coast). All the patients had cerebral CT and biological check up.

Results: Patients' mean age was 42 years. The sex ratio was 1.3. The coma was deep (score of Glasgow = 8) in 77.5 % of the cases (n=110). Comas with fever represented 54.9 % of the cases (n=78) and comas without fever represented 45,1 % of the cases (n = 64). HIV - AIDS was associated in 39.4 % (n = 56) of the cases to coma with fever. In 69.7 % (n=99) of the cases, an etiology was found at cerebral CT scan. These causes were represented by cerebral toxoplasmosis (33.8 %; n = 48), ischemic stroke (16.9 % n = 24), brain haematoma (7.8 % n=11), brain tumors (6.4 % n = 9) and subarachnoid hemorrhage (2.1 % n = 3). In 30.3 % (n = 43) of cases the cerebral CT scan was normal and biological etiologic finding were neuro-malaria (10.6 % n = 15), cerebral tuberculosis (4.6 % n = 6), neuro-meningeal cryptococcosis (2.8 % n = 4), diabetic coma (3.5 % n = 5), hyperthyroidism (1.4 % n = 2) and toxic coma (2.1 % n = 3). In 5.7 % of the cases (n = 8) no cause was found.

Conclusion: in Sub-Saharan Africa, the etiologies of comas with fever are dominated by toxoplasmosis cerebral and those without fever are dominated by stroke.

[*] Tél: +22505190002 email: kngoran@yahoo.fr

The cerebral CT scan helped to find more than the 2/3 of the causes of non traumatic comas and it also helped an immediate management of the pathology.

Keywords: cerebral CT scan; comas; etiology; toxoplasmosis cerebral; stroke.

INTRODUCTION

Coma is the symptom of a brain suffering. It is defined as an irreversible impairment of vigilance by stimuli. Coma is related to a global neuronal dysfunction or more particularly to the dysfunction of the activating reticular substance situated in the brain stem [1]. It is a frequent reason for hospitalization in the emergency and intensive care units [2,3]. Its therapeutic management is carried out in an emergency context and is dominated by the preservation of vital functions which can fail. Specific treatments must be instituted immediately (according to the causes). These causes are many: they can be traumatic and non traumatic (toxic, vascular, metabolic, deficient, infectious, epileptic [1]. Traumatic comas are often made evident by the context (road accident, assault or fall) [4]. It is not the case of non traumatic comas which raise a true etiologic problem. This etiologic check up consists of a complete and systematic clinical examination including a detailed interrogation of people around in search for indications, a repeated somatic and neurologic examination of different reflexes in particular. Biology brings complementary information to causes but it must be targeted according to the most probable diagnostic hypothesis [1]. Medical imaging remains essential in this etiological check up [3,5]. In the past it was represented by the standard radiography of the cranium and was confined to the search or the confirmation of syndromes of intracranial hypertension. The advent of the Computed Tomography Scan (CTS) and the Magnetic Resonance Imaging (MRI) help assess the morphologic and functional aspect of the encephalon today [5]. The Magnetic Resonance Imaging (MRI) is not always available in emergency. The Computed Tomography Scan (CTS) constitutes the only imaging means which can be performed in emergency in the case of comas.

The aim of our work was to make clear the importance of the Computed Tomography Scan in the management of non traumatic comas and to list the main etiologies of non traumatic comas in sub-Saharan Africa.

PATIENTS AND METHODS

We carried out over a four-year-period (from January 2006 to December 2010) a descriptive and analytic retrospective study about non traumatic comas in sub-Saharan Africa. This study was carried out in the departments of Radiology, Emergency, Neurology and Intensive care of Yopougon University Hospital (Abidjan-Côte d'Ivoire, West Africa). It concerned the analysis of records and encephalic Computed Tomography images of 142 patients hospitalized in the Neurology, Neurosurgery, Medical emergency and Intensive care departments of Yopougon University Hospital for non traumatic coma. All the Computed Tomography Scans were performed in the Radiology department of Yopougon University Hospital. In all the patients the Computed Tomography Scans consisted of a Cranioencephalic acquisition in spontaneous contrast. In patients whose images were suspected of infection or tumors the Computed Tomography Scans consisted of a second acquisition after the injection of iodine-containing contrast media. The interpretation of the Computed Tomography Scans was performed by senior radiologists in parenchymatous and bone windows after three-dimensional reconstructions. We consulted patients' records in their departments of origin for the etiologic confirmation, the clinical elements and the results of laboratory tests including HIV serology, blood smear, lumbar puncture, the search for toxic substance in the blood and urines, glycemia etc. Some clinical elements like temperature had been useful to divide comas into two groups: febrile comas (temperature above 37.5°C) and non febrile comas (temperature below or equal to 37°C). The importance or the depth of the coma based on Glasgow scale [6] has been taken down in patients' records. In some records Glasgow scale was not used for the classification of the coma [7,8]. The coma was classified in stage I,II,III and IV. In other records the coma was classified in obnubilation or mental confusion, light coma, deep coma, irreversible coma. For harmony we used in our study the following equivalence: Glasgow score = 13: stage I coma (mental confusion or obnubilation); Glasgow score = 9-10: stage II coma (light coma), Glasgow score = 8-6: stage III coma (deep coma), Glasgow score ≤ 5: stage IV coma (irreversible coma). The working out of the questionnaire and the capture of results were performed by using the software program SPHINX and the processing and analysis of data were performed by using Microsoft World 2007 an Excel 2007.

RESULTS

Patients' mean age was 42 years with extremes of 28 and 84 years. Non traumatic comas were predominant in the age bracket between 30 and 40 years (35.9%) followed by the one between 40 and 50 years (21.1%). Age groups from 20 to 30 years, 50 to 60 years, 60 to 70 years and from 80 to 90 years accounted for respectively 2.1%, 19.7%, 17.6%, 2.8% and 0.7% of cases (Table I). The sex ratio was 1.3 (81 men and 61 women). In 77.5% of cases (n=110) it was about deep coma (Glasgow score=7; Table II). Light comas (Glasgow score between 9 and 13) and irreversible comas (Glasgow score ≤5) were found in a respective proportion of 20.4% (n=29) and 2.1% (n=3). There was no case of mental confusion (Glasgow score >13). Comas (Table III) were associated with fever (febrile coma) in 54.9% of cases (n=78). Among these patients 39.4% (n=56) presented a positive HIV serology (all types considered, that is HIV1 and / or HIV 2). The twenty-two other febrile coma cases presented a negative HIV serology. The cases of coma without fever (non febrile coma) were 64 in number (45.1% of cases). Only one among them presented a positive HIV serology. The etiologies of febrile comas were the following:

- brain toxoplasmosis: 33.8% (n=48)
- meningoencephalitis: 2.8% (n=4).
- cerebral malaria or pernicious malaria: 10.6% (n=15)
- tuberculous meningitis: 4.2% (n=6)
- neuromeningeal cryptococcosis: 2.8% (n=4)

Table I. Distribution of comatose patients according to the age group

Age group	Number	Percentage
20-30	3	2.11
31-40	51	35.91
41-50	30	21.13
51-60	28	19.72
61-70	25	17.61
71-80	4	2.82
81-90	1	0.70
Total	142	100

More than half (59%) of patients were less than 50 years old.

Table II. Distribution of patients according to the depth of the coma (Glasgow scale)

Glasgow score	Number	Percentage
≥ 13	-	-
9-13	29	20.4
8-6	110	77.5
≤ 5	3	2.1
Total	142	100

Deep coma (stage III coma) was predominant (77.5%).

Table III. Distribution of patients according to the type and etiologies of the coma

Type of coma	**Number**	**Percentage**
Febrile	**78**	**54.9**
Brain toxoplasmosis	48	33.8
meningoencephalitis	4	33.8
cerebral malaria	15	10.6
tuberculous meningitis		
neuromeningeal Cryptococcus	4	2.8
no cause	1	0.7
Non febrile	**64**	**45.1**
ischemic stroke	24	16.9
intra-parenchymatous hematoma	11	7.9
meningeal hemorrhage	3	2.1
brain tumor	9	6.4
diabetic coma	5	3.5
hyperthyroidism	2	1.4
poison induced coma	3	2.1
no cause	7	4.9
Total	**142**	**100**

Febrile comas (54.9%),were more common than non febrile comas (45.1%). Brain toxoplasmosis (33.8%) was the most common cause of coma followed by ischemic strokes (16.9%).

Among febrile comas, no cause was found in one case (0.7%)
The etiologies of non febrile comas were the following:

- ischemic stroke: 16.9% (n=24)
- intra parenchymatous hematoma: 7.8% (n=11)
- meningeal hemorrhage: 2.1% (n=3)
- brain tumor: 6.4% (n=9)
- diabetic coma: 3.5% (n=5)
- hyperthyroidism: 1.4% (n=2)
- poison induced coma: 2.1% (n=3)

Among non febrile comas, no cause was found in 7 cases (4.9%).

The Cranio-encephalic Computed Tomography Scan (Table IV) demonstrated in 69.7% of cases (n=99) a coma etiology. It was about:

- comas of infectious (figure 1) particularly parasitic etiologies (toxoplasmosis) were predominant (33.8%; n=48) and they appeared in the form of brain abscesses. Thirty-five (72.9%) of them were at the collection stage and thirteen (27.1%) at the pre-suppurative stage.
- comas of vascular etiologies. Ischemic strokes (figure 2) accounted for 16.9% (n=24) of our cases, followed by parenchymatous hematomas (7.8% n=11) and by meningeal hemorrhage (2.1% n=3)
- comas of tumor etiology (6.4% n=9). Brain tumors were benign (3 cases of meningioma), primary malignant (2 cases of glioblastoma, 1 case of astrocytoma, 1 case of lymphoma) and secondary malignant (2 cases of brain metastasis, one of which was of mammary origin and the other of hepatic origin).

Table IV. Distribution of causes of coma according to the results of the Computed Tomography Scan

Results of the CT Scan	Number	Percentage
Pathological Computed Tomography Scan	**99**	**69.7**
brain toxoplasmosis	48	33.8
meningoencephalitis	4	2.8
ischemic stroke	24	16.9
intra-cerebral hematoma	11	7.7
meningeal hemorrhage	3	2.1
brain tumor	9	6.4
Normal Computed Tomography Scan	**43**	**30.3**
cerebral malaria	15	10.6
tuberculous meningitis	6	4.2
neuromeningeal Cryptococcus	4	2.8
diabetic coma	5	3.5
hyperthyroidism	2	1.4
Poison induced coma	3	2.1
nothing to report	8	5.7
Total	**142**	**100**

The Cranioencephalic Computed Tomography Scan demonstrated 69.7% of causes of non traumatic coma.

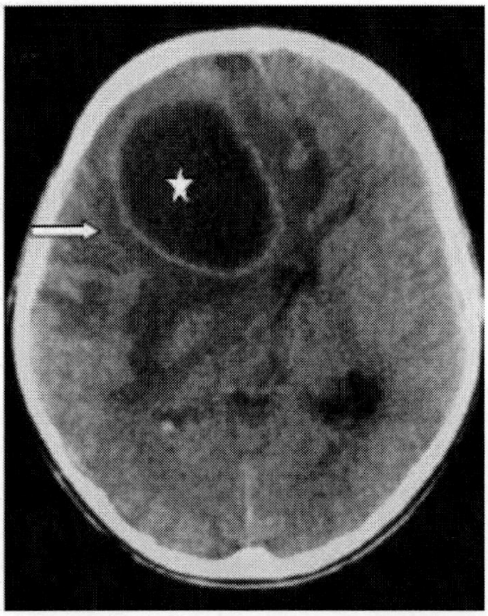

Figure 1. Cranioencephalic Computed Tomography scan after the injection of iodine-containing contrast media in axial section and in parenchymatous windows in a comatose patient. Brain abscess (star) located in the right frontal lobe surrounded by an edema (arrow) with an important mass effect on the median structures (sub-falcorial and temporal herniation).

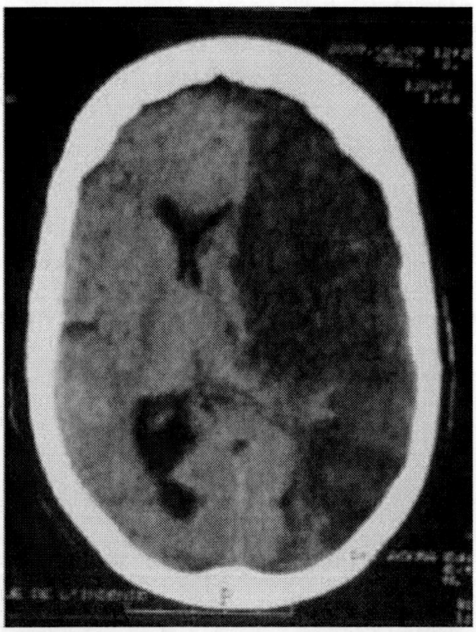

Figure 2. Encephalic Computed Tomography Scan in axial section with the injection of iodine-containing contrast media in a comatose patient. Total Sylvian brain infarction with an important edema and hernia.

The cerebral Computed Tomography Scan was normal in 30.3% of cases (n=43). The causes which were not found on the Computed Tomography Scan but which were demonstrated by biology were the following:

- infectious comas: cerebral malaria (10.6% n=15) tuberculous meningitis (4.6% n=6), neuromeningeal cryptococcosis (2.8% n=4).
- diabetic comas (3.5% n=5), hyperthyroidism (1.4% n=2).
- poison induced comas (2.1% n=3)

In 5.7% of cases (n=8) neither the Computed Tomography Scan nor biology helped find the cause of the coma.

COMMENT

1. Types of Coma

Classically, comas divide up into two groups: traumatic and non traumatic comas [4]. This distinction is justified by the mode of management of the coma. Traumatic comas which are also called surgical comas are managed in a neurosurgical unit whereas non traumatic comas which are also called medical comas are treated in emergency or intensive care or medical neurology units.

Non traumatic comas have several etiologies. They are usually classified in six sub-groups. It's about metabolic, vascular, deficient, infectious, epileptic and poisoned induced comas. But these etiologies appear many. They need to be regrouped in two entities to permit an effective management of coma in emergency. Authors in the Western countries [9] oppose structural or lesional comas (which express the existence of a lesion at the level of intracranial structures) to non structural or non lesional comas (absence of lesion at the level of intracranial structures).

In sub-Saharan and tropical regions, the classification of non traumatic comas uses only two entities too. It's about comas associated with fever (febrile comas) and comas without fever (non febrile comas). In our work febrile comas were more common (54.9%) than non febrile comas (45.1%). This tropical classification of coma appears simple and practical. It permits to quickly associate an antibiotics or anti-parasitic based treatment with urgent measures aiming at the preservation of the patient's vital functions (after biological samplings). But this classification is subject to reservations in relation with the definition of coma or its physiopathology. Any coma stipulates the presence of important intra-cerebral lesions (which can be visualized or not at the medical imaging) with a bilateral affection of the ascending activating reticular formation of the brain stem, the neurophysiologic support of vigilance. The consequence of these lesions is the loss of the waking reaction but also cause vegetative disorders justifying the quick implementation of intensive care measures. Among these vegetative disorders, the loss of thermal regulation control can be noticed. Thus all coma can be accompanied by hyperthermia or hypothermia. It is the case of hypoglycemic coma which is accompanied by hypothermia [4]. The interrogation of people around or the consultation of the record (health record in our context) must help know the chronology between the onset of the coma and the appearance of fever. It is then important to know that febrile coma is the coma which is preceded by fever or which is febrile at once. Any fever which appears 24 or 48 hours after the occurrence of the coma should be attributed to a vegetative disorder related to the coma itself. It can also be about an added malarial attack (if we are in an endemic area of malaria)(justifying the performance of a blood smear) and not at the origin of the coma.

Taking the temperature is in sub-Saharan Africa and in the tropical regions an important step in the clinical examination of a comatose patient. This step must be performed at the same time as all the measures aiming at the preservation of the vital functions of the comatose patient.

2. Causes of Coma

2.1. Febrile Comas

Febrile comas are called infectious comas in the developed countries where they seem to be too little common not to justify a full entity [9,10]. In sub-Saharan Africa, febrile comas are the most common causes of non traumatic comas. They accounted for more than half of cases in our work (54.9%). The reason for this high frequency may be its association with HIV/AIDS which is rife in an endemic way in sub-Saharan Africa.

In our study 56 patients out of 78 (71.8%) suffering from febrile coma were HIV positive. Febrile comas affect younger subjects (patients' mean age in our work was 42 years with a predominance (35.9%) of the age bracket between 30 and 40 years). This could justify the fact that brain toxoplasmosis, an opportunistic disease of HIV/AIDS was the most common disease (33.8%) in our work. It is followed by cerebral malaria (10.6%), tuberculosis meningitis (4.8%), neuromeningeal cryptococcosis (2.8%) and meningoencephalitis (2.8%).

2.1.1. Brain Toxoplasmosis

Brain toxoplasmosis is all the manifestations engendered by the location of toxoplasma Gondii in the central nervous system. Usually, it is about an endogenous reactivation of an old disease following an immunosuppression [11,12], in particular AIDS during which it constitutes the most common opportunistic disease. It can account for up to 60% of neurologic manifestations during the course of AIDS and it can be the inaugural event and a discovery circumstance [13,14]. Brain toxoplasmosis is often manifested by disorders of consciousness developing from mental confusion into irreversible coma. In our study, patients were often met at the stage where the coma was serious (Glasgow score=7) because in Africa consultations were always late. Because of socioeconomic problems [11,15] and mystic beliefs [16] African patients arrive at a reference hospital at the late stage of the disease. But the stage of the coma also depends on the severity of the disease which is materialized by the size of the lesion or the edema which accompanies it (figure 1) with the effects of masses on the median structures (temporal herniation). The location of the abscess is also incriminated. A location of the posterior fossa or bi-thalamic doesn't have the same manifestation as a location of the parietal lobe. The multiciplicity of lesions in the same patient is part of the gravity factors of coma. Patients presenting bilateral lesions both supra and sub tentorial arrive at the hospital with a serious coma right away.

In our study 2 patients out of the 3 who had a Glasgow score ≤5 had brain toxoplasmosis.

In addition to the coma the temperature was high at once. Avodé and al. [11] revealed a fever superior to 39.5°C. This fever often precedes the coma from one to two weeks. It is often a factor of diagnostic delay because it is the source of self medication or diagnostic error. In sub-Saharan Africa, because of the fact that malaria is endemic, any fever is first treated as an attack of malaria. It is faced with this persistent fever despite the treatment of the attack of malaria that the consultation and the infectious check up are performed [1]. The third cardinal symptom guiding to brain toxoplasmosis is the associated motor deficiency. More often it is about hemiplegia (100% in our case) of progressive installation. Other signs such as epileptic seizures, visual disorders and the syndrome of cranial hypertension can be met in brain toxoplasmosis [8].

2.1.2. Clear Fluid Meningitis

Viral meningoencephalitis, tuberculous meningitis and neuromeningeal cryptococcosis are regrouped under the term clear fluid meningitis. It is about diseases which are relatively common. They accounted for 9.8% (n=4) of causes of non traumatic comas in our study . They are manifested by a coma with variable gravity associated with a fever of little importance, generally below 39°C. The main sign is the meningeal syndrome (present in 100% of cases in our study). The clinical examination must help eliminate any focal sign in order to perform a lumbar puncture which permits to make a positive diagnosis. In the study carried out by Soumaré and al. in 2005 in Dakar [19], clear fluid meningitis was the cause of coma in 28% of cases. The etiologies were represented by neuromeningeal Cryptococcosis (29 cases, that is 63%) and neuromeningeal tuberculosis (5cases, that is 10%). In 26% of cases no etiology was found. All the patients were HIV positive. In our study 11 patients out of 14 were HIV positive (that is 78.6% of cases). In Mali Minta and al. in 2010 [20] revealed 75% of HIV positive patients out of 31 cases with clear fluid meningitis. The etiologies were the following: neuromeningeal Cryptococcosis (9 cases), viral meningoencephalitis (9 cases) neuromeningeal tuberculosis (2 cases), decapitated bacterial meningitis (1 case) and brain tumor (1 case). In 10 cases the cause was unspecified.

The prognosis was bad. The global lethality was 63% (29 cases) reaching 83% in the absence of identified cause [19].

2.1.3. Cerebral Malaria

Cerebral malaria is also called pernicious malaria or serious malaria. Cerebral malaria is the neurologic manifestation of falciparum malaria which is rife in an endemic way in sub-Saharan Africa [21,22]. Cerebral malaria is manifested by convulsions followed by coma [23]. In our study it accounted for 10.6% of cases (n=15) of coma. The prognosis of comas related to cerebral malaria is bad. The high mortality rate of this form of malaria is also related to a problem of management, since patients often arrive at the hospital too late despite the availability of an effective treatment. Different hypotheses have been advanced to explain the pathogenesis of cerebral malaria: disorder of vascular permeability, immunopathologic mechanism, intravascular coagulation and mechanical obstruction of micro-vessels, metabolic effects. Each of these hypotheses had had its hour of glory, its defenders and its therapeutic implications more or less justified. At present the hypothesis which is considered as more probable admits a central role for the inter-capillary sequestration .of infected erythrocytes by cytoadherence to different endothelial molecules. Coma is probably due to a lack of brain oxygenation caused by a reduction of blood flow and a partial mechanical obstruction by the reduction of the use of oxygen by the brain tissue resulting from an inhibition of the mitochondrial functions by the inflammatory process. In both cases local hypoxia is the determining factor of coma. When hypoxia coincides either with hypoglycemia or with an accumulation of lactic acid (as a result of a hyper parasitemia in capillaries following the sequestration), or an accumulation of glutamate (as a result of a reduction of the consumption of glutamate by the astrocytes induced by nitric oxide, NO), the neurologic damages are always more important than hypoxia alone [2]. The mortality of serious malaria is not just related to coma. It is particularly related to be associated with multi-visceral failures [24].

2.2. Non febrile Comas

Non febrile comas represent all the disorders of consciousness except any previous or present episode of fever. They consist of comas of vascular, metabolic, tumor and poison induced origin. In our study non febrile comas accounted for 45.1% of cases distributed among comas of vascular origin (26.8%), tumor origin (6.4%), metabolic origin (4.9%) and poison induced origin (2.1%). In seven cases (4.9%),no cause was found. They might be cases of coma related to brain anoxemia of various origins and which are difficult to prove [9,25] or cases of infarction of the brain stem non identified on the Computed Tomography Scan [9].

2.2.1. Comas of Vascular Origin

It is about strokes. In the literature [26] strokes are distributed among ischemic strokes (80%), hemorrhagic accidents (15%) and deep venous thrombosis (5%). In our study, vascular comas (figure2 and 3) were represented by ischemic strokes (16.9%), intra-parenchymatous hematoma (7.8%) and meningeal hemorrhage (2.1%). We did not find any deep venous thrombosis because deep venous thrombosis which is also called acute thrombophlebitis occurs in the young woman using contraceptives and its most common clinical manifestation is more headaches than coma. Vascular comas are comas which occur suddenly accompanied with a motor deficiency and more often on a known hypertension predisposition. The depth of the coma is variable and is more often in relation with the compressive underlying edema. Intra-parenchymatous hematomas are bad prognosis in comparison with the coma related to brain infarction [26] because of the risk of flooding of brain ventricles by hemorrhage, the cause of an acute hydrocephaly [9].

2.2.2. Brain Tumors

Brain tumors are causes of non febrile comas, which is constantly reported in the literature [9,28]. They accounted for 6.4% (n=9) of all the cases of coma included in our study. Their mechanism is in relation with the edema which accompanies the tumor process or its sub-tentorial or intraventricular location [9]. Their diagnosis is made in the context of headaches, coma of progressive installation associated with sensorimotor deficiencies. They occur more often in the subject who is more than 50 years old with impairment of the general condition except brain lymphoma which occurs in the young subject, on a HIV/AIDS predisposition [18]. In our study brain tumors were benign (3 cases of meningioma), primary malignant (2 cases of glioblastoma, 1 case of astrocytoma, 1 case of lymphoma) and secondary malignant (2 cases of brain metastasis: one of mammary origin and the other of hepatic origin). If the prognosis is at once bad for secondary malignant tumors and to a lesser extent for lymphomas, the prognosis of glioblastomas and astrocytomas depends on the precocity of the diagnosis. Meningiomas don't put at risk the prognosis for life if they are operated but there is a known risk of relapse.

2.2.3. Metabolic Comas

In our study metabolic comas are represented by diabetic comas and hyperthyroidism. We noticed two cases of hypoglycemia in known diabetic patients who presented non febrile coma with focal signs. The presence of asymmetric motor reaction (focal signs) during hypoglycemic comas is described in the literature [29]. It is the same for the association of several causes of coma in the same patient [9]. This is the reason why even if the clinical signs of hypoglycemic coma (diabetic predisposition, quiet coma, sweating, hypothermia, bilateral Babinski sign) are clear and confirmed by the quantity determination of glycemia [4,29,30], it is necessary to carry out other complementary explorations. The acute thyroid crisis classically occurs after an intercurrent event worsening a pre-existent hyperthyroidism. The clinical picture is characterized by tachyarrhythmia, confusion or delirium and digestive disorders. An early diagnosis and an emergency treatment are essential to reduce the high morbidity and mortality rate of this disease [31]. It is about a very uncommon cause of comas, accounting for 1.4% of cases (n=2) of coma encountered in our work.

2.2.4. Toxic Comas

In our study poison induced comas were the most uncommon etiologies. They accounted for 2.1% of cases (n=3). Adulterated alcohol was the toxic substance which was found and it had no suicide objective. The diagnosis of alcohol poisoning was guided by the patient's breath and the interrogation of people around. The aim of paraclinical explorations was to eliminate another cause associated with the alcohol predisposition such as cranial traumatism, hypoglycemia or meningeal hemorrhage [4]. The prognosis was always good with a recovery of consciousness in the 48 hours without sequels.

3. Contribution of Cranioencephalic Computed Tomography Scan

Any coma, from its mechanism deserves the investigation of the Cranioencephalic chamber. According to Dubas [9], two types of mechanism can cause coma.

- a diffuse pain of the brain cortex (functional depression, hypomatabolism of all the neurons of the brain cortex), during poison induced or metabolic comas, meningeal hemorrhage or meningitis, generalized epilepsy seizure (or after a generalized epilepsy seizure)
- a pain of the ascending activating reticular formation which can be compressed by an expansive process (stroke, tumor etc…….) of the posterior fossa or during a temporal herniation (hernia of the internal part of the temporal lobe between the tentorium cerebelli and the mesencephalon), this expansive process itself secondary to a hemispheric expansive process. The temporal herniation compresses the perimesencephalic venous network causing hemorrhagic lesions (petechias and/or small hematomas of the mesencephalon). The ascending activating reticular formation can also be injured directly (ischemic stroke or hemorrhage, tumor or infectious process of the brain stem). During poison induced or metabolic comas the neuronal hypometabolism concerns the brain cortex but it also concerns the ascending activating reticular formation. It is then useful to look for encephalic lesions during any coma. This research is only possible with medical imaging and

particularly the encephalic Computed Tomography Scan [25]. In our study the Computed Tomography Scan helped bring out 2/3 of causes of comas (69.7% of cases). The pathologies found on the Computed Tomography Scan were febrile comas (36.6%) followed by non febrile comas (33.1%). In fact the Computed Tomography Scan contributes to the diagnosis of all the cases of brain abscess, comas of vascular and tumor origin. These different etiologies are called structural or lesional causes in the literature [9,10,29]. Nevertheless a reservation must be expressed about the infarctions of the brain stem [26] which is not visualized by the Computed Tomography Scan. In our study the cause of a case of non febrile coma was neither found with biology nor on the Computed Tomography Scan. The Magnetic Resonance Imaging (MRI) should be recommended in this case. But the problem of availability and the waiting period for patients to have the examination is still raised.

The Computed Tomography Scan is not a contributory factor in metabolic and poison induced comas [2]. All the patients who had had a diabetic coma, a coma in relation with hyperthyroidism and alcohol poisoning had had a normal Computed Tomography Scan. Nevertheless the Computed Tomography Scan had had the merit to eliminate another associated cause or a factor which could make the symptoms worse. The contribution of Computed Tomography Scan to febrile comas must be nuanced . In 1/3 of cases (32%, n=25) the Computed Tomography Scan didn't bring any information. The contrast enhancement of subarachnoid spaces associated with a brain swelling can guide to meningoencephalitis without prejudging its etiology. Nevertheless any anomaly of temporal density guides to herpetic meningoencephalitis with an immediate specific management [9]. In our study the Computed Tomography Scan demonstrated 4 cases of meningoencephalitis. Besides cerebral malaria, neuromeningeal Cryptococcosis and tuberculous meningitis do not modify the Computed tomography aspect of the encephalon. But the lumbar puncture which allows to make their positive diagnosis (apart from cerebral malaria) should be preceded by an encephalic Computed Tomography Scan. In one case, neither the Computed Tomography Scan nor biology could make the diagnosis of febrile coma in our study.

CONCLUSION

The causes of comas are similar in sub-Saharan Africa as they are everywhere throughout the world. Nevertheless their proportions are different and their prognoses depend on the patient's predisposition and on the co-morbidity. Because of malaria and HIV/AIDS double endemic in sub-Saharan Africa, infectious comas have reached very important proportions and should change from a simple group of cause to a full entity. Thus the distinction of coma in 2 entities (febrile coma and non febrile coma) in sub-Saharan Africa appears practical and constitutes a rapid method of management of comas in emergency. The Computed Tomography Scan plays an important role in the management of these comas and should immediately follow the measures of preservation of vital functions and the clinical examination. In our study it demonstrated 2/3 of all the causes of comas and 2/3 of causes of febrile comas. Contrary to the studies carried out in the western countries, infectious comas are the most common in sub-African Africa and toxic comas the most uncommon.

REFERENCES

[1] Martin P, Piroelle B. Comas. EMC (Elsevier Masson SAS), *Médecine d'urgence*, 25-110-A-20, 2007.

[2] Adukauskiene D, Budryte B, Karpec D. *Coma: etiology, diagnosis, and treatment. Medicina* (Kaunas). 2008;44(10) : 812-9.

[3] De Castro P. Patients with alteration of consciousness in the emergency department. *An. Sist. Sanit. Navar.* 2008; 31: 87-97.

[4] Danziger N, Alamowitch S. Neurologie, édition De Boeck, Paris 2004, 448 pages.

[5] Tshibanda L, Vanhaudenhuyse A, Bruno MA, Boly M, Soddu A et Col. Imagerie par résonance magnétique chez des patients survivant à un coma. *Rev Med Liege* 2009 ; 63 : 42-8.

[6] Teasdale G, Jennett B. Assessment of coma and impaired consciousness. A practical scale. *Lancet* 1974; 2:81-4.

[7] Ledoux D, Piret S, Boveroux P, Bruno M-A, Vanhaudenhuyse A, Damas P, Moonen G, Laureys S. Les échelles d'évaluation des états de conscience altérée. *Réanimation* (2008) 17, 695—701.

[8] Wijdicks EF. Clinical scales for comatose patients: the Glasgow Coma Scale in historical context and the new FOUR Score. *Rev. Neurol. Dis*. 2006; 3(3):109-17.

[9] Frédéric Dubas. Coma non traumatique. *Rev Prat* 2007 ; 5 7 : 321-6.

[10] Young GB Coma. *Ann. N. Y. Acad. Sci.* 2009; 1157:32-47.

[11] Avodé DG, Adjien KC, Houinato DS, Sounhin M, Adoukonou TA. Toxoplasmose cérébrale en milieu hospitalier à Cotonou (Benin). *African journal of neurological sciences* 2005 ; 24 : 48-54.

[12] Finielz P, Chuet C, Ramdame M, Guiserix J. Traitement de la toxoplasmose cérébrale au cours du SIDA par le Cotrimoxazole. *Presse Med.* 1995; 24, 19: 917.

[13] Price RW. Neurological complications of HIV infection. *Lancet* 1996, 17; 348 (9025): 445-52.

[14] Millogo A, Ki-Zerbo GA, Traore W, Sawadogo AB, Ouedraogo I, Pelhini M. Sérologie toxoplasmique chez les patients VIH positifs et suspects de toxoplasmose cérébrale au centre hospitalier de Bobo-Dioulasso (Burkina-Faso). *Bull Soc. Pathol. Exot.* 2000; 93 (1): 17-9.

[15] Kouna Ndouongo P, Millogo A, Siéméfo Kamgang FP, Assengone-Zeh Y. Aspects épidémiologiques et évolutifs des accidents vasculaires au centre hospitalier de Libreville (Gabon). *African journal of neurological sciences* 2007 ; 26 (2) : 12-7.

[16] Ndri-Yoman T, Mahassadi AK, Touré A, Bathaix FMY, Kissi HY, Bangoura DA, Doffou S, Attia AK. Croyances mystiques et syndrome de l'intestin irritable chez le patient noir africain. *J. Afr. hépatol. Gastroentérol.* 2010 ; 4 :104-9.

[17] Eholié SP, Ehui E, Adou-Bryn K, Kouamé KE, Tanon A, Kakou A, Bissagnené E et Kadio A. Paludisme grave de l'adulte autochtone à Abidjan (Côte d'Ivoire). *Bull Soc. Pathol. Exot.,* 2004, 97, 5, 340-344.

[18] Lacroix C. Manifestations neurologiques du VIH. Encycl Méd Chir (Elsevier, Paris), *Neurologie*, 17-051-B-10, 1999, 11 p.

[19] Soumaré M, Seydi M, Ndour CT, Dieng Y, Ngom-Faye NF, Fall N, Diop BM. Les méningites à liquide clair chez les patients infectés par le VIH à Dakar. *Bull Soc. Pathol. Exot.* 2005 ; 98(2) :104-7.

[20] Minta DK, Dembele M, Kaya AS, Sidibe AT, Coulibaly I, Mieret S, Diallo B, Traore A, Ba B, Sidibe AF, Diallo D, Traore AK and Traore HA. Les méningites lymphocytaires à Bamako, Mali. *Mali Med.* 2010; 25 (2):17-22.

[21] Beghdadi W, Porcherie A, Schneider BS, Peronet R, Louis J, Mécher S. Rôle de l'histamine et des récepteurs histaminiques dans la pathogenèse du paludisme. *Médecine/Sciences* 2009 ; 25 : 377-81.

[22] Dondorp AM, Fanello CI, Hendriksen IC, Gomes E, Seni A and al. Artesunate versus quinine in the treatment of severe falciparum malaria in African children (AQUAMAT): an open-label, randomised trial. *Lancet* 2010; 376(9753):1647-57.

[23] Severov MV, Comolet T, Magassouba N, Barry M, Diallo AAS. Epidémiologie du paludisme grave chez l'adulte à Conakry (guinée) : implication pour la stratification et le contrôle. *Med. Afr. Noire* 2000, 47 (2) : 65-71.

[24] Bruneel F. Prise en charge du paludisme grave de l'adulte. *La Lettre de l'infectiologue* 2006 ; 21(6): 263-276.

[25] Boveroux P, Kirsch M, Boly M, Massion P, Sadzot B, Lambermont B, Lancellotti P, Piret S, Damas P, Damas F, Moonen G, Laureys S, Ledoux D. Neurologic prognosis assessment in postanoxic encephalopathy. *Reanimation* (2008) 17, 613-61.

[26] Haute Autorité Sanitaire, Service des bonnes pratiques professionnelles : Accident vasculaire cérébral : prise en charge précoce (alerte, phase pré hospitalière, phase hospitalière initiale, indications de la thrombolyse). Paris : HAS ; 2009.

[27] Crassard I, Ameri A, Rougemont D et Bousser MG. Thromboses veineuses cérébrales. Encycl Méd Chir (Elsevier SAS, Paris), *Neurologie* [17-046-R-10].

[28] Ciais JF, Memran N. Coordination médicale à domicile pour la prise en charge palliative d'un patient porteur d'une tumeur cérébrale. Médecine Palliative : Soins de Support – Accompagnement. *Ethique* 2005 ; 4 : 29-32.

[29] Perkin GD. Neurologie: Manuel et atlas. Édition de Boeck; Paris 2002: 320 pages.

[30] Orban JC, Lena D, Bonciu M, Grimaud D, Ichai C. Complications métaboliques aiguës du diabète. *Urgence pratique* 2007 ; 83 :13-7.

[31] Léger A. Hyperthyroïdie et hypothyroïdie. EMC (Elsevier Masson SAS), Urgences [24-132-A-10].

In: Comas and Syncope: Causes, Prevention and Treatment ISBN 978-1-62100-603-9
Editors: Eduardo Silva and Gabriel Cruz. ©2012 Nova Science Publishers, Inc.

Chapter 7

THE ROLE OF NEUROIMAGING AND GENETICS IN COMATOSE PEDIATRIC PATIENTS

E. Brichtova

Clinic of Pediatric Surgery, Orthopaedics and Traumatology,
Brno Faculty Hospital, Czech Republic

ABSTRACT

Coma is defined as a profound or deep state of unconsciousness. Comatose patient cannot voluntarily react with the external environment and mostly possess specificaly limited reflex reactivity to external stimuli depending on coma severity. Different levels of coma severity can be determined by neurological examitation based on brainstem reflexes and level of unresponsiveness to external stimuli. Coma severity and further prognosis depend upon the nature and extent of brain dysfunction, determined primarily by causative etiologic factors. Traumatic brain injuries are one of the most common causes of coma. Coma is commonly a result of non-traumatic intracerebral bleeding, ischaemic stroke or inflammatory insult with or without brain swelling. Generally, coma is also result of inadequate oxygen or glucose supply to brain tissue and a great variety of drug and poisons can directly cause a coma too. Initial diagnosis and treatment in comatose patient should start simultaneously. History still remains the basal key to the diagnosis. Complete clinical neurologic assessment, Glasgow Coma Score and clinical physical examination together with electrocardiography, brain computer tomography or magnetic resonance imaging scan, electroencephalography, blood, urine and cerebrospinal fluid laboratory examination should follow in close connection. Immediate treatment is focused to life-threatening issues, as open airway, spontaneous breathing, heart beat and blood pressure, blood glucose level, serum osmolarity and acid-base balance of internal environment. Comatose patients are treated in intensive care units and usually require full life support until their situation resolves in either way.The prognosis and outcome of comatose patient depends on the coma cause, the extent and severity of brain damage, the possibility to correct or normalised affected brain function and comatose state duration. Repeated neurologic assessment is needed to monitoring the coma level. A coma possessing defined clinical features that lasts longer than four weeks is usually reclassified as a persistent vegetative state, sometimes remaining for years. Comas resulting from drug overdosage or poisonings have the good recovery capacity if immediate and proper medical care is received. Generally comas resulting from traumatic brain injuries tend to have a slightly better recovery rate compared to comas related to illness. However comatose patient may never regain consciousness or be permanently seriously disabled.

Vegetative state lasting more than one year has a minimal chance to regain full consciousness. Predicitve factors also include electrophysiology, biochemical and genetic testing regarding evoked potential responses, S 100 protein and apolipoprotein E genotype. Age related specifics need to be considered in comatose pediatric patients.

INTRODUCTION

Coma or unconsciousness is a partial or complete loss of the central nervous system brain function, it means the ability to accept and process information at higher than reflective level. Consciousness is a state of mind, which is based on alertness (vigilance), orientation, thought and attention.

The causes of coma are multifactorial and can be divided into structural and functional disability of brainstem activation system and diffuse disability of neuronal metabolism. Coma therefore may be caused by focal structural lesion with bilateral haemispheric damage; by focal brainstem lesion (especially of the upper pons); or by metabolic origin with suppression of the ascendent reticular activating system. Several neurological disorders and general medical conditions are capable to cause coma.

Structural lesions leading to unconsciousness are in the first place of vascular origin - they are all types of intracranial hemorrhage (epidural, subdural, intracerebral, intraventricular and subarachnoideal hemorrhage) (Figures 1, 2, 3, 4, 5) and focal brain ischemia (on the basis of thrombosis, vasospasm, embolism, or vascular abnormalities) (Figure 6).

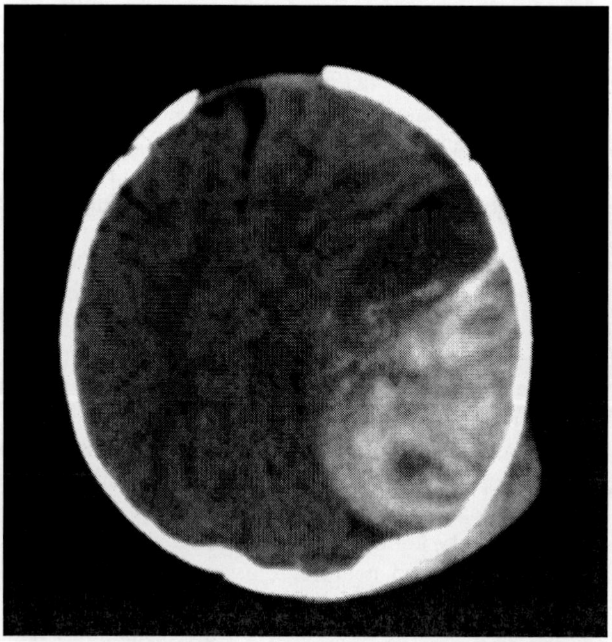

Figure 1. Epidural hemorrhage.

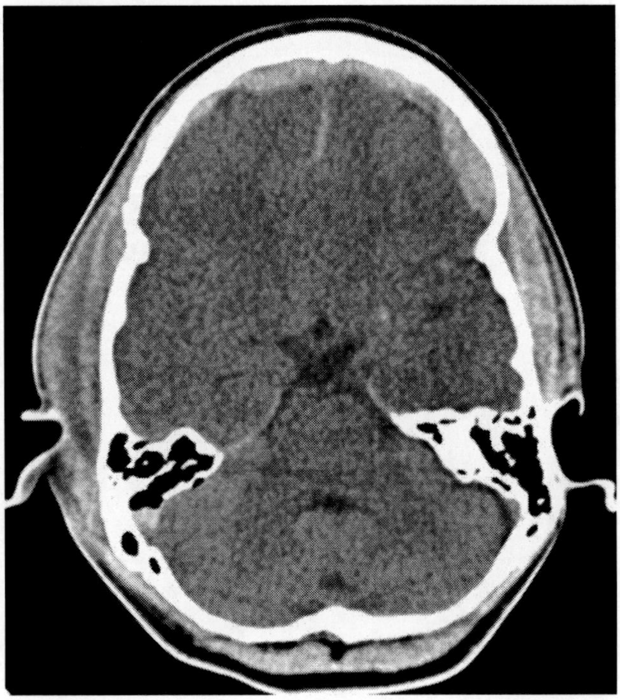

Figure 2. Subdural hemorrhage.

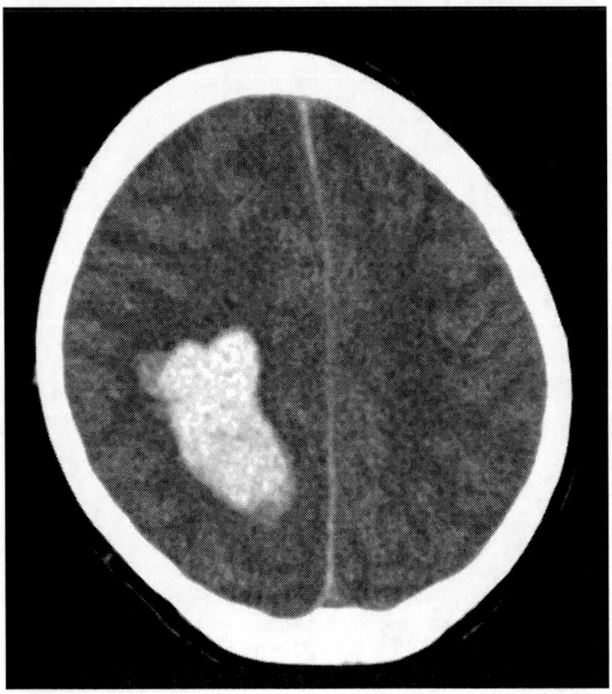

Figure 3. Intracerebral hemorrhage.

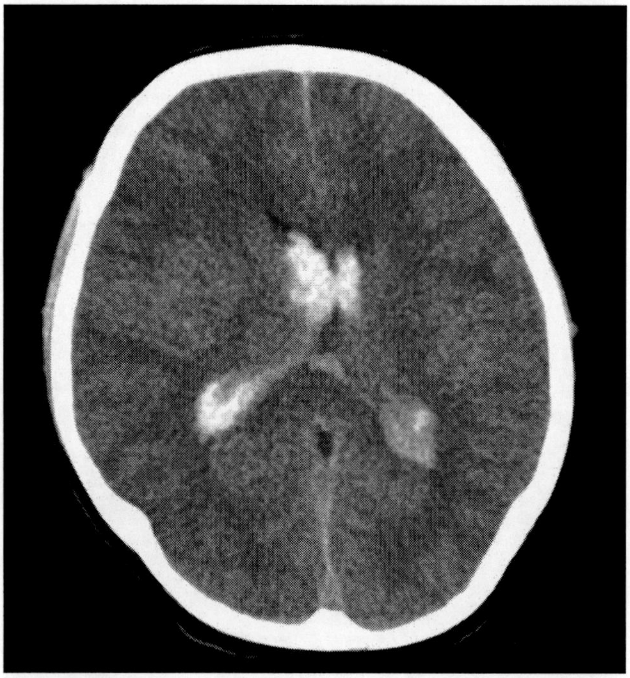

Figure 4. Intraventricular hemorrhage.

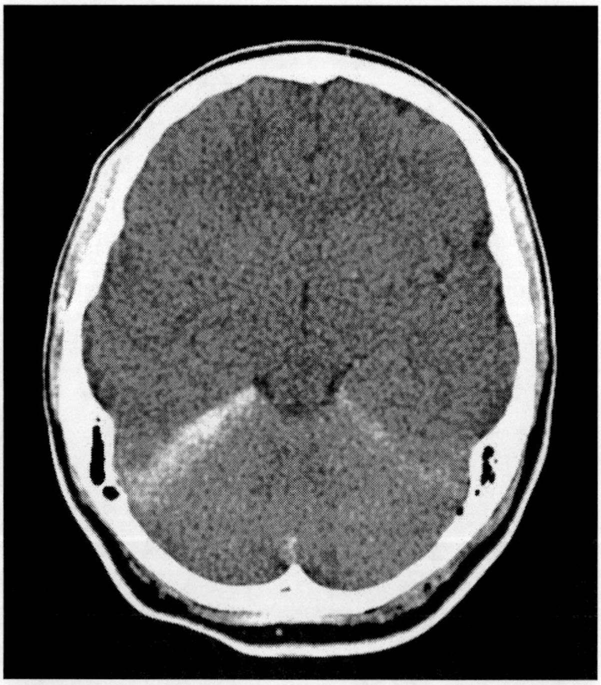

Figure 5. Subarachnoideal traumatic hemorrhage.

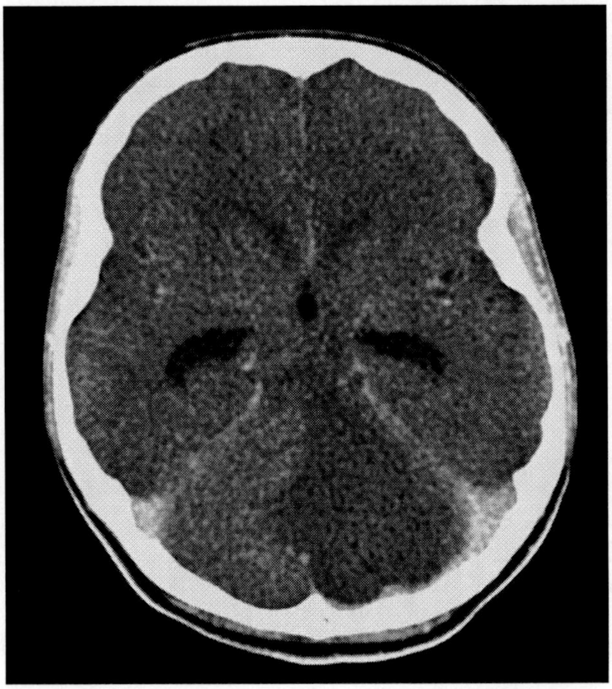

Figure 6. Cerebellar ischemia.

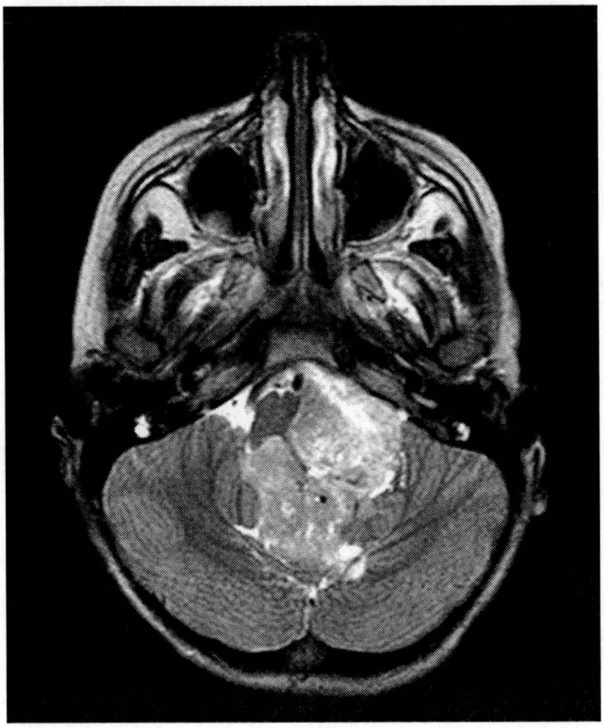

Figure 7A. Ependymoma.

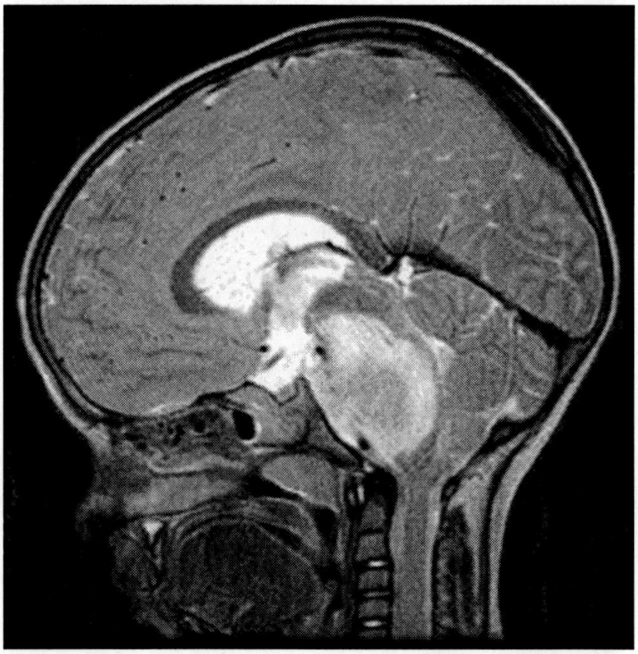

Figure 7B. Brainstem glioma.

Structural changes are also of infectious origin (meningitis, encephalitis, brain abscesses) or caused by cerebral tumors (Figures 7A, 7B). Another unconsciousness causes represent intoxications caused by sedatives, hypnotics, anesthetics, psychotropic substances, alcohol, carbonmonoxide and numerous other drugs and poisons. Metabolic causes of coma are associated with disabilities resulting from the neuronal dysmetabolism on the basis of diffuse ischemia, hypoglycemia, hypo-and hyperthermia, endocrinopathy, with a deficit coenzymes such as thiamine or nicotinic acid deficiency. Another cause is the neuronal membrane functional disability induced by hypo-and hyperosmolarity, acid-base dysbalance, seizure or neurotrauma. The multifactorial causes comprise organ failure - septic, liver and uraemic encephalopathy and multiorgan failure. According to the unconsciousness duration, it is also possible to classify coma on the short-term and prolonged. Short-term unconsciousness takes a few seconds to minutes and in children is most commonly associated with head trauma and cerebral concussion. All patients suffering from cerebral concussion should undergo neurological examination, skull radiography and surgical hospitalisation. Patiens with prolonged unconsciousness are hospitalized at intensive care unit and undergo brain computer tomography (CT) or magnetic resonance imaging (MRI).

The role of neuroimaging in patients with sudden onset of impaired consciousness is irreplaceable and allows diagnosis of significant cerebral lesions. The most common cause of unconsciousness in children are craniocerebral injuries. Mortality as a result of head injury is 5 times higher than childhood leukemia mortality and 18 times higher than mortality from brain tumors [1]. Head injury is the leading cause of death in children over one year, in children under one year of age occupies third place. The greatest mortality shows the youngest age categories of children under 6 years. The incidence of head injury has increasing tendency with age in boys, in girls decreases with age. The most common causes of craniocerebral injuries are falls and transport accidents [1,2]. CT is still sovereign method in the diagnosis of acute bleeding in the first 24 hours. Cerebral hemorrhage is usually clearly visible immediately. For subarachnoid hemorrhage the blood is mainly found in the basal cisterns and central sulcus. The most important CT benefit is a reliable diagnosis and differentiation of intracranial hemorrhage. In comparative pathologic studies, however, CT showed little sensitivity in the detection of cortical contusion without bleeding, primary brainstem injury and in the detection of diffuse axonal injury [3-5]. Diffuse axonal injury together with cortical contusion is considered the most important and most frequent cause of morbidity in patients [6,7].

It is now clear that the CT alone does not provide sufficient information for reliable and accurate determination of the degree and extent of brain injury. MRI is considered as a very sensitive diagnostic method for brain injury without concomitant bleeding by many authors [3, 5, 6, 8]. Brain injury is caused by two different mechanisms. Surface brain contusion is caused by the impact of the brain to the inner surface of the skull bones, rarely to the dura mater folds. This generates cortical brain contusion.

The second type of injury does not depend on direct contact with brain and skull. This injury is caused by "scissors" mechanism of rotational very fast acceleration and/or deceleration. These changes are most pronounced at the interface tissues with different density, at the junction of white and gray matter. The changes are usually diffuse and bilateral. Diffuse axonal injury is present in up to 50% of patients with intraaxial brain injury [5]. Typical clinical manifestation is immediate and deep coma. Diffuse axonal injury can rarely be present without other types of brain injury [8].

Another large group of acute comas results from spontaneous intracranial bleeding of non-traumatic origin. Blood collection from damaged vessel wall is localized in the surrounding cerebral tissue or in the cerebrospinal fluid spaces. According the bleeding localisation therefore the spontaneous intracranial bleeding can be devided to intracerebral and subarachnoideal hemorrhage (SAH). SAH is a major cause of mortality and permanent morbidity, mortality is approximately 50% [9,10]. The most common cause of spontaneous SAH is intracranial aneurysm rupture. Early SAH diagnosis and source of bleeding finding is crucial to achieve good therapeutic results. Digital subtraction angiography (DSA) is currently the gold standard for examination of patients with SAH or intracranial aneurysm suspected [11]. Less invasive angiography diagnostics based on multidetector computer tomography (MDCT) and magnetic resonance technology replaced DSA at a number of sites [6,12].

MDCT is a non-invasive imaging method. With the advantage it can be done immediately after detection of SAH in the initial native CT scan [13]. MDCT permits significant reduction of investigation time compared to DSA, even in non-cooperating patients, and allows image reconstruction in any plane. Clear advantages of MRI angiography are the absence of ionizing radiation and any need of contrast substance application or catheterization. Otherwise requires patient cooperation and the need to respect the MRI contraindications [4].

MRI angiography is limited by the artifacts in blood flow slowing, tiny size aneurysms or subacute SAH, also there are significant motion artifacts in non-cooperating subjects [7,14]. Arterio-venous malformation (AVM) (Figure 8) is much less frequent cause of intracerebral bleeding, the same as cavernous angioma.

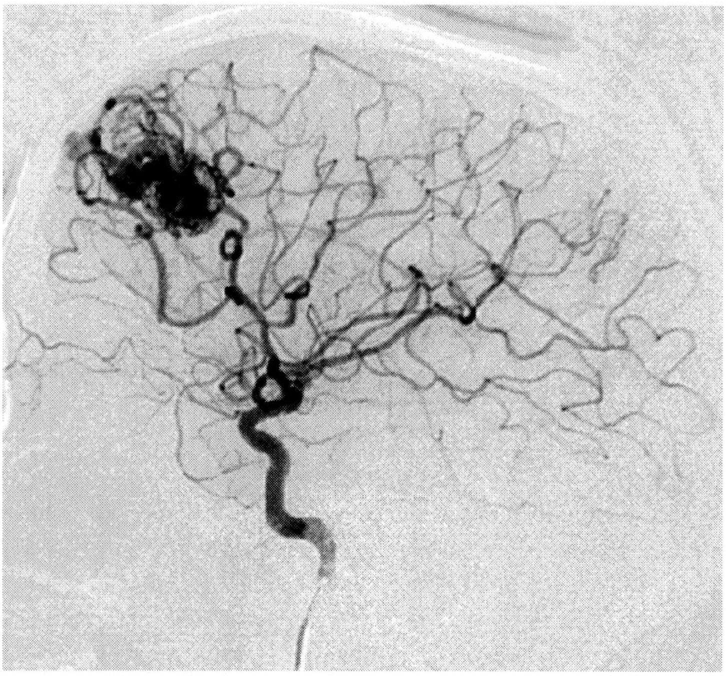

Figure 8. Arteriovenous malformation.

The exact distinction between these so-called "benign" bleeding and bleeding into the cerebral tumor may be difficult. Frequently there is a bleeding in vascular tumors, such as gliomas or hemangioblastoma. "Benign" bleeding is usually characterized by a homogeneous appearance with hemosiderin rim visible on MRI scans. On the contrary, tumor has usually heterogeneous composition, application of contrast agent can feed the tumor center. Often the final decision depends on the CT image development, the spontaneous benign bleeding leads to decreasing both the size and collateral edema. Hemorrhage in the lateral parts of the cortex is always suspected of AVM hemorrhage or metastases. Nowadays we can meet more often with bleeding from arteries damaged by cocaine and amphetamine in drugs addicted.

This is confirmed that CT is usually unable to demonstrate the diffuse axonal injury changes in the most affected areas, those are the white matter of brain hemispheres and the area of corpus callosum. MRI diffuse axonal injury findings yet has no defined impact on the way of treatment, but allows precise injury classification and identification of the degree of brain injury. At the same time also predict potential clinical development and explain the possible discrepancy between CT findings and severe clinical status. Patients suffering from SAH in poor clinical condition primarily undergo MDCT, patients in good clinical condition profit from MRI examination. If the source of bleeding is not proven and the findings are unclear, than DSA is performed.

The principles of comatose patient management in childhood are similar to those in adults, as the coma aethiology is associated with structural brain changes and metabolic dysfunction too. Despite this there are several age specific factors that should be taken into consideration evaluating comatose children [15,16]. The Glasgow Coma Scale (GCS) is commonly used to classify the coma severity. There are many difficulties using the GCS in pediatric patients, especially regarding the verbal scale, which is clearly inappropriate to young children capabilities, there are also difficulties with the motor response tasks [15,17].

The modified GCS was developed for reliable assessment of coma in childhood [15,18]. Modified GCS is applicable for children under and above 5 years of age, both motor and verbal responses are adequately included in that scale.

The enhanced neuronal plasticity in children paradoxically possess a dangerous predisposition for severe neuronal hypoxia. Increased neural plasticity in childhood may explain the increased susceptibility for brain damage in young children [15,19]. The developing brain is also more vulnerable to drugs that impair neuronal activity, such as glutamate antagonists and gamma aminobutyric acid (GABA) agonist sedatives [15,20].

Intracranial pressure (ICP) increase is an important issue in comatose pediatric patients. There are reports suggestive for an age dependent susceptibility to brain swelling in children [15, 21-23]. These studies point to a need for intensive monitoring of ICP in comatose children. An open fontanel may keep an increase in ICP from causing cerebral herniations and attenuate the symptom of headache. ICP measurement and monitoring has become routine practice in pediatric neurosurgery and is often indicated in children with traumatic brain injury treated at intensive care units. ICP assessment is usually performed via ventricular cannulation or intracerebral devices.

Neurophysiological methods have been applied for assessment of brain function in the management of severe brain injury in children, especially electroencephalography (EEG) and evoked potentials, namely somatosensory and auditory modalities (SSEP, BAEP). EEG is essential in the diagnosis of non convulsive status epilepticus (NCSE), which is commonly seen in comatose pediatric patients due to enhanced seizure susceptibility in the early years of life. Transcranial color coded doppler ultrasonography (TCCD) have became usefull tool for assessment of intracerebral circulation.

The different course of illness and different response to treatment in patients with craniocerebral injury is probably genetically determined and supported by different degree of neuroplasticity, the ability of formation the neural synapses and different immunological response to cerebral insult. Therefore, among other, the relationship of apolipoprotein E (ApoE) genotype to the clinical course and consequences of craniocerebral injury has been studied [1,24]. ApoE is produced by astrocytes and microglia and plays a role in transporting lipids in the nervous system by influencing lipoprotein uptake [25, 26]. This transport system provides nerve cells by the cholesterol and phospholipids needed for the maintenance and repair of neuronal membranes, dendritic remodeling, and synaptogenesis [27]. Generally, ApoE is thought to be responsible for the transportation of lipids, namely, cholesterol, within the brain, maintaining structural integrity of the microtubule within the neuron, assisting with neural transmission, and plays a role in immunology response to cerebral trauma [28]. Astrocytes contained in the blood–brain barrier produce ApoE to injured neurons, so that ApoE may play an isoform-specific role in determining the initial response and the subsequent development to acute brain injury [29-31]. The impact of injury to the central nervous system is in part to be modulated by the different isoforms of ApoE. ApoE exists in three isoforms named E2, E3, and E4. The gene associated with ApoE influences the construction and regeneration of the neuronal cytoskeleton in allele-specificmanner: ApoE $\varepsilon 2/\varepsilon 2$ may be neuroprotective, whereas ApoE $\varepsilon 4/\varepsilon 4$ may be neurodestructive [25]. The neuroprotective effect of E3 allele was observed in experimental conditions. The way of action of ApoE is not fully understood. In experimental conditions, the neurodegenerative properties of ApoE $\varepsilon 4$ allele can be attenuated by calcium channel blockers [32].

Evidence from previous studies indicates that ApoE plays a defined role in the response of the brain to injury and implicates the possession of the Apo ε4 allele as a genetic determinant in recovery from head injury. Published studies demonstrated the correlation between ApoE ε4 expression and the unfavorable outcome of traumatic brain injury in adult patients, with higher risk of long-term posttraumatic coma [33], more than five times higher probability of unconsciousness for more than 7 days [29], worsening of cognitive functions, learning disturbances, and worsening of memory [34]. More recent evidence indicates an association with poor outcome after intracerebral hemorrhage. ApoE gene polymorphism also influences the risk of hemorrhage in cerebral amyloid angiopathy [35], and expression of ApoE ε4 allele was associated with more severe hypoxic damage and with higher cerebral contusion index [33]. The presence of ApoE ε4 genotype is also mentioned in relation with increased risk of late onset Alzheimer's disease and cardiovascular disorders [27, 26, 36-38].

Results of recent studies show significant association of ApoE genotype and the course and prognosis of traumatic brain injury also in children. Pediatric patients possessing the ApoE epsilon 4 genotype were more likely to have severe clinical symptomatology and unfavorable neurological outcome after traumatic brain injury compared to significantly better outcome with other ApoE genotype. These data can be of clinical use by predicting the outcome of traumatic brain injury in children. Implementation of the ApoE genotyping to a routine clinical management of pediatric head trauma can improve the initial assessment and treatment of traumatic brain injury in children. Also in less severe head injuries [39-41], the ApoE ε4 genotype can indicate possible neuropsychological deficit and thus allow to establish early neuropsychological and social care.

Another protein biomarker used for the detection of cerebral injury and outcome prediction is S100B protein. The serum level of S100B was confirmed as a clinically valuable indicator of the severity of cerebral injury and was proposed as an effective predictor of risk outcome. The S100 proteins are a family of dimeric cytosolic calcium-binding proteins made up of an alpha and a beta isomer. S100 proteins are found in the astroglial and Schwann cells and have also been found in certain tumours [42]. Elevation in serum S100B protein levels have ben reported to reflect cerebral injury and increased permeability of the blood-brain barrier. The initial S100B serum levels can provide diagnostic and prognostic indications of the cerebral structural lesion at an early stage of serious cerebral injury, especially in comatose pediatric patients with non-conclusive initial neuroimaging findings. Serial measurements of S100B levels can also monitor secondary events and contribute to better case assessment and treatment. The severity of secondary brain injury is expected to have stronger relationship with outcome than primary injury [42].

Conclusion

Coma may be caused by focal structural lesion with bilateral haemispheric damage; by focal brainstem lesion or by metabolic origin with suppression of the ascendent reticular activating system. Coma severity and further prognosis depend upon the nature and extent of brain dysfunction, determined primarily by causative etiologic factors. Traumatic brain injuries are one of the most common causes of coma and head injury is the leading cause of death in children over one year. Due to the high mortality of comatose children suffering from craniocerebral injury it is necessary to provide immediate and correct diagnosis especially by choosing proper diagnostic tools and establish adequate early therapeutic measures. Comatose children differ from comatose adults in several aspects. Age related specifics need to be considered in comatose pediatric patients diagnosis and treatment. The role of neuroimaging methods is prominent. The principles of comatose patient management in childhood comprise several age specific factors, especially susceptibility to brain swelling, vulnerability to excitatory stimulation loss, raised intracranial pressure symptoms and epileptic seizures susceptibility. Implementation of the ApoE genotyping to a routine clinical management of pediatric head trauma can improve the initial assessment and treatment of traumatic brain injury in children. We should also pay great attention to prevent accidents primarily.

References

[1] Brichtova E. Kraniocerebrální poranění v dětském věku. Praha, Triton 2008.

[2] Brichtova E. Analýza souboru pacientů s kraniocerebrálním poraněním léčených na Pracovišti dětské medicíny FN Brno v období let 2000–2007. *Cesk. Slov. Neurol.* N, 2008;71(4): 466–471.

[3] Gentry LR, Godersky JC, Thompson B. MR imaging of head trauma. *Amer. J. Roentgenol.* 1988; 150: 663–672.

[4] Marquez MP, Sant LM, Roig AJB, et al. Utility of nuclear magnetic resonance in the diagnosis of diff use axonal injury in head trauma. *Med. Intensiv.* 1996; 20(1): 1–5.

[5] Osborn AG. Diagnostic Neuroradiology. St. Louis, Mosby 1997.

[6] Teasdale G, Teasdale E, Hadley D. Computed tomographic and magnetic resonance imaging classifi cation of head injury. *J. Neurotrauma* 1992; 9(Suppl.1): 249–257.

[7] Wilson JTL, Hadley DM, Wiedmann KD, Teasdale GM. Neuropsychological consequences of two patterns of brain damane shown by MRI in survivors of severe head injury. *J. Neurol. Neurosurg. Psychiat.* 1995; 59(3): 328–331.

[8] Colohan ART, Oyesiku NM. Moderate head injury: an overview. *J. Neurotrauma* 1992; 9(Suppl.1): 259–263.

[9] Jayaraman MV, Mayo–Smith WW, Tung GA, et al. Detection of intracranial aneurysms: Multi-detector row CT angiography compared with DSA. *Radiology* 2004; 230: 510–518.

[10] Krajina A, Náhlovský J. Endovaskulární léčba aneuryzmat mozkových tepen. In Krajina A, Peregrin JH, a kol. Intervenční radiologie. Hradec Králové: *Olga Čermáková*, 2005; 405–424.

[11] Fendrych P. MR angiografie v diagnostice aneuryzmat intrakraniálních tepen. Kam s ní? *Čes. Radiol.* 1998; 52(Suppl.): 44–46.

[12] Tipper G, U-King-Im JM, Price SJ, et al. Detection and evaluation of intracranial aneurysms with 16 – row multislice CT angiography. *Clin. Radiol.* 2005; 60: 565–572.

[13] U-King-Im JM, Koo B, Trivedi RA, et al. Current diagnostic approaches to subarachnoid haermorrhage. *Europ. Radiol.* 2005; 15: 1135–1147.

[14] Young N, Dorsch NW, Kingston RJ. Pitfalls in the use of spiral CT for identification of intracranial aneurysm. *Neuroradiology* 1999; 41: 93–99.

[15] Bragatti, JA. Considerations for the Pediatric Coma Patient: Not Just Small Adults. *Pediatr. Health.* 2010, 4(6), 581-589.

[16] Seshia SS, Bingham WT, Griebel RW: Coma in childhood. In: ***Disorders of Consciousness. Handbook of Clinical Neurology, Volume 90 (3rd series)***. Young GB, Wijdicks EFM (Eds). Elsevier, London, 2008, 329–352.

[17] Kirkham FJ, Newton CR, Whitehouse W: Paediatric coma scales. ***Dev. Med. Child. Neurol.*** 2008, 50, 267–274.

[18] Tatman A, Warren A, Williams A, Powell JE, Whitehouse W: Development of a modified paediatric coma scale in intensive care practice. ***Arch. Dis. Child.*** 1997, 77, 519–521.

[19] Samatovicz RA: Genetics and brain injury: apolipoprotein E. ***J. Head Trauma Rehabil.*** 2000, 15, 869–874.

[20] McDonald JW, Johnston MV: Physiological and pathophysiological roles of excitatory amino acids during central nervous system development. ***Brain Res. Brain Res. Rev.*** 1990, 15, 41–70.

[21] Zimmerman RA, Bilaniuk LT, Bruce DA ***et al.*** : Computed tomography of pediatric head trauma: acute cerebral swelling. ***Radiology***. 1978, 126, 403–440.

[22] Bruce DA, Alavi A, Bilaniuk L ***et al.*** : Diffuse cerebral swelling following head injuries in children: the syndrome of 'malignant brain oedema'. ***J. Neurosurg.*** 1981, 54, 170–178.

[23] Obrist WD, Gennarelli TA, Segawa H ***et al.*** : Relation of cerebral blood flow to neurological status and outcome in head-injured patients. ***J. Neurosurg.*** 1979, 51, 292–300.

[24] Brichtova E, Kozak L. Apolipoprotein E genotype and traumatic brain injury in children – association with neurological outcome. *Childs Nerv. Syst.*, 2008; 24(3): 349–356.

[25] Kerr ME, Kraus M. Genetics and the central nervous system: apolipoprotein E and brain injury. *AACN Clin Issues.* 1998, 9:524–530.

[26] Samatovicz RA. Genetics and brain injury: apolipoprotein E. *Head Trauma Rehabil.* 2000, 15:869–874.

[27] Graham DI, Horsburgh K, Nicoll JA, Teasdale GM. Apolipoprotein E and the response of the brain to injury. *Acta Neurochir.* 1999, Suppl 73(X):89–92.

[28] Nathoo N, Chetty R, van Dellen JR, Barnett GH. Genetic vulnerability following traumatic brain injury: the role of apolipoprotein E. *Mol. Pathol.* 2003, 56:132–136.

[29] Friedman G, Froom P, Sazbon L, Grinblatt I, Shochina M, Tsenter J,Babaey S, Yehuda B, Groswasser Z. Apolipoprotein E epsilon4 genotype predicts a poor outcome in survivors of traumatic brain injury. *Neurology.* 1999, 15:244–248.

[30] Laskowitz DT, Horburgh K, Roses AD. Apolipoprotein E and the CNS response to injury. *J. Cereb. Blood Flow. Metab.* 1998, 18:465–471.

[31] Lynch JR, Pineda JA, Morgan D, Zhang L, Warner DS, Benveniste H, Laskowitz DT. Apolipoprotein E affects the central nervous system response to injury and development of cerebral edema. *Ann. Neurol.* 2002, 51:113–117.

[32] Veinbergs I, Everson A, Sagara Y, Masliah E. Neurotoxic effects of apolipoprotein E4 are mediated via dysregulation of calcium homeostasis. *J. Neurosci. Res.* 2002, 67(3):379–387.

[33] Smith C, Graham DI, Murray L, Nicoll JAR. Association of APOE polymorphisms and pathological features in traumatic brain injury. *Neuropathol. Appl. Neurobiol.* 2002, 28(2):151–152.

[34] Seliger G, Lichtman SW, Polsky T, Riley J, Tycko B, Marder K. The effect of apolipoprotein E on short-term recovery from head injury. *Neurology.* 1997, 48(3 Suppl):28001.

[35] Schiefermeier M, Kollegger H, Madl C, Schwarz C, Holzer M, Kofler J, Sterz F. Apolipoprotein E polymorphism: survival and neurological outcome after cardiopulmonary resuscitation. *Stroke*. 2000, 31(9):2068–2073.

[36] Horsburgh K,McCarronMO,White F, Nicoll JA. The role of apolipoprotein E in Alzheimer's disease, acute brain injury and cerebrovascular disease: evidence of common mechanisms and utility of animal models. *Neurobiol. Aging.* 2000, 21(2):245–255.

[37] Saunders AM, Strittmatter WJ, Schmechel D, George-Hyslop PH, Pericak-Vance MA, Joo SH, Rosi BL, Gusella JF, Crapper- MacLachlan DR, Alberts MJ. Association of apolipoprotien E allele epsilon 4 with late-onset familiar and sporadic Alzheimer's disease. *Neurology.* 1003, 43(8):1467–1472.

[38] Ueki A, Kawano M, Namba Y, Kawakami M, Ikeda K. A high frequency of apolipoprotien E4 isoprotein in Japanese patients with late-onset nonfamiliar Alzheimer's disease. *Neurosci. Lett.* 1993, 163(2):166–168.

[39] Chamelian L, Reis M, Feinstein A. Six-month recovery from mild to moderate traumatic brain injury: the role of APOEepsilon 4 allele. *Brain.* 2004, 127:2621–2628.

[40] Collie A, Maruff P, Falleti M. APOE influences on neuropsychological function after mild head injury: within-person comparisons. *Neurology.* 2004, 63:2460.

[41] Sorbi S, Nacmias N, Piacentini S, Repice A, Latorraca S, Forleo P, Amaducci L. ApoE as a prognostic factor for posttraumatic coma. *Nat. Med.* 1995, 1(2):135–137.

[42] Zurek J, Bartlova L, Marek L, Fedora M. Serum S100B protein as a molecular marker of severity in traumatic brain injury in children. 2010, 73/106(1), 37-44.

In: Comas and Syncope: Causes, Prevention and Treatment ISBN 978-1-62100-603-9
Editors: Eduardo Silva and Gabriel Cruz.

Chapter 8

INACCURACY AND MISJUDGED FACTORS OF GLASGOW COMA SCALE SCORES

Jun Namiki*

Department of Emergency and Critical Care Medicine,
Keio University School of Medicine, Tokyo, Japan

ABSTRACT

Traumatic brain injury and other neurological disorders are primarily assessed using the Glasgow Coma Scale (GCS) to evaluate consciousness. However, GCS score assignment is far from sufficiently reliable for correct assessment, especially with inexperienced users. In this chapter, we reveal what factors of the GCS are misjudged when assessed by inexperienced medical personnel. We analyzed GCS eye, verbal, and motor response (EVM) scoring profiles conducted by postgraduate year 1 junior residents (n = 94) before they began residency in specific departments. GCS assessment was tested using a video simulation that portrayed mock patients with eight different levels of consciousness that are frequently encountered in trauma patients at the emergency department. On average, 26% ± 18% of examinees failed to provide the correct EVM profiles for the eight selected consciousness levels. Primary misjudged GCS factors belonged to two categories: the assessment of "confused conversation (verbal score of 4)", and the assessment of "withdrawal motor response (motor score of 4)". Additional instruction regarding the specific misjudged factors described in this chapter may help inexperienced medical personnel improve the reliability of GCS score assignment to casualties or patients with consciousness disturbance.

1. INTRODUCTION

The Glasgow Coma Scale (GCS) was developed as an objective measure of the level of consciousness [1] and originally developed to assess coma and impaired consciousness level after severe head injury [1, 2]. The GCS comprises assessments of eye (E), verbal (V), and

* Telephone: 81-3-3225-1323, fax: 81-3-3353-2232, E-mail: namiki@sc.itc.keio.ac.jp

motor (M) responses graded on scales of 4, 5, and 6, respectively (Table 1) [2]. The total GCS score is derived from the sum of the scores allocated to these three components, and ranges from 15 to 3 [3]. Although the total GCS score is widely used to express level of consciousness, the EVM profile should also be recorded along with the total GCS score, because it has been shown that two groups with the same composite score but different EVM profiles may have significantly different outcomes [4].

Traumatic brain injury and other neurological disorders are primarily assessed using the GCS to evaluate consciousness. However, GCS score assignment is far from sufficiently reliable for correct assessment, especially with inexperienced users. Misinterpretation or poor knowledge of the GCS definition would lead to misjudgment of consciousness levels. Rather than attempting to teach all definitions of all 15 scales (consisting of the 4 scales of E, 5 scales of V, and 6 scales of M), education and study focused on specific individual components in which errors result in the misjudgment of GCS scoring would most likely be the effective approach to GCS education. In this chapter, we reveal what factors of the GCS are misjudged when assessed by inexperienced medical personnel [5].

Table 1. Glasgow Coma Scale [2, 28].

Component	Level	Score
Eye opening	Spontaneous	E4
	To speech	3
	To pain	2
	None	1
Best verbal response	Oriented	V5
	Confused conversation	4
	Inappropriate words	3
	Incomprehensible sounds	2
	None	1
Best motor response	Obeys commands	M6
	Localizes pain	5
	Withdrawal (normal flexion)	4
	Abnormal flexion (decorticate)	3
	Extension (decerebrate)	2
	None	1

2. GCS and Emergency Patients

Appropriate triage of a large number of patients with minor head injury is crucial at the emergency department (ED) as well as at the field in management decisions such as the requirement of a computed tomography (CT) and transfer to a neurosurgical facility. For example, ED patients involving mild head injury with a GCS score of 14 or 13 should be distinguished from those with a GCS score of 15 in order to properly identify patients requiring CT [6, 7]. In some European countries, all patients with a GCS below 14 are centralized in hospitals with neurosurgical facilities [8]. In clinical practice undertriage of patients with minor head injury is sometimes critical. For example, in recommendations of the National Health Service in England and Wales, a patient with GCS scores less than 15 on initial assessment is a criterion for referral to a hospital ED by community health services and clinics [9]. As military concerns or in case of sport players, minor casualties with a GCS score of 14, especially the EVM profile of E4V4M6, who may be able to walk and keep on their mission, should be rested and observed by medical personnel because of medical meanings and an adequate performance of a task.

In some circumstances, deeply comatose patients with severe head injury may be triaged as out of emergency surgical intervention. Moreover, it is often difficult to initially realize a brain injury patient from among various trauma patients. In the context of the triage based on severity and treatment urgency of casualties, the assessment of physiological condition is a first priority. Thus, traumatic brain injury is primarily assessed by a level of consciousness, and it has been recommended that GCS is used.

3. Reliability of GCS Scores

The reliability of the GCS is of great relevance to clinicians in the fields of neurosurgery [10-13] and critical care [14-16], as well as to physicians in the ED [17-21]. The GCS is considered to be fairly reliably measured by trained medical personal [22]. Some investigators have reported that the GCS has good interrater reliability in the ED [19, 20]. However, it has been reported that the reliability of the GCS is insufficient in clinical practice [10, 12, 18, 23]. Because it is impossible to retrospectively determine whether or not a GCS score assigned to a patient on his/her medical record was correctly determined, investigations into GCS reliability have been conducted as prospective observational studies, with analysis of GCS scoring agreement among medical personnel [11-13, 15-20]. Although it has been pointed out that some specific patient conditions, including intubation [24-27] and changes in consciousness level during the early period after insult [24, 28], adversely affect GCS reliability, specific errors in GCS assessment have never been disclosed, except for the poor distinction between “abnormal” and “normal” flexion [13]. Thus, information which could aid the effort to better improve the accuracy of the GCS score is very sparse.

4. The Most Common GCS Presentations of ED Patients

We reviewed the chart database of consecutive patients who were transported to the ED by ambulance during a 3-year period (from April 2003 to March 2006) [5]. Patients were categorized into those with or without trauma. Initial assigned, GCS-based EVM scores were obtained from a total of 10,354 patient records including 4,066 cases with trauma. The evaluations of the EVM scores were performed by emergency physicians or post graduate year (PGY) 1 or 2 junior residents under the supervision of emergency physicians. Analyzed patient records did not include the following: patients initially referred to the ward doctors on duty; patients under 6 years old, in whom it is difficult to assess the consciousness level using the GCS; cases intubated prior to ED admission, who were given the non-numerical designation of “T” for the verbal GCS score [24, 25, 27]; and cases with missing database records regarding their consciousness level.

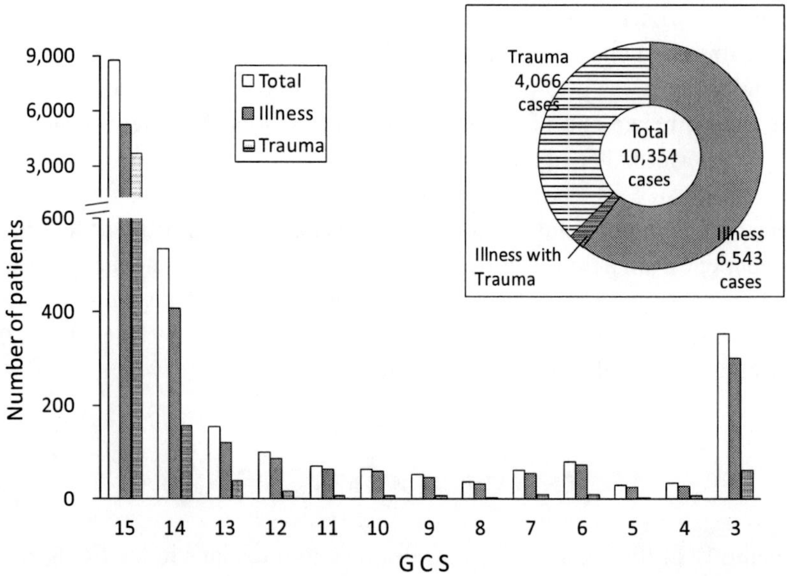

Figure 1. The total GCS scores representing initial consciousness level of ED patients. The total GCS scores derived from the sum of the EVM scores were obtained from consecutive patients (n=10,354) who were transported to the ED by ambulance during a 3-year period. The main reason for calling of the ambulance were grouped into illness or trauma (inset).

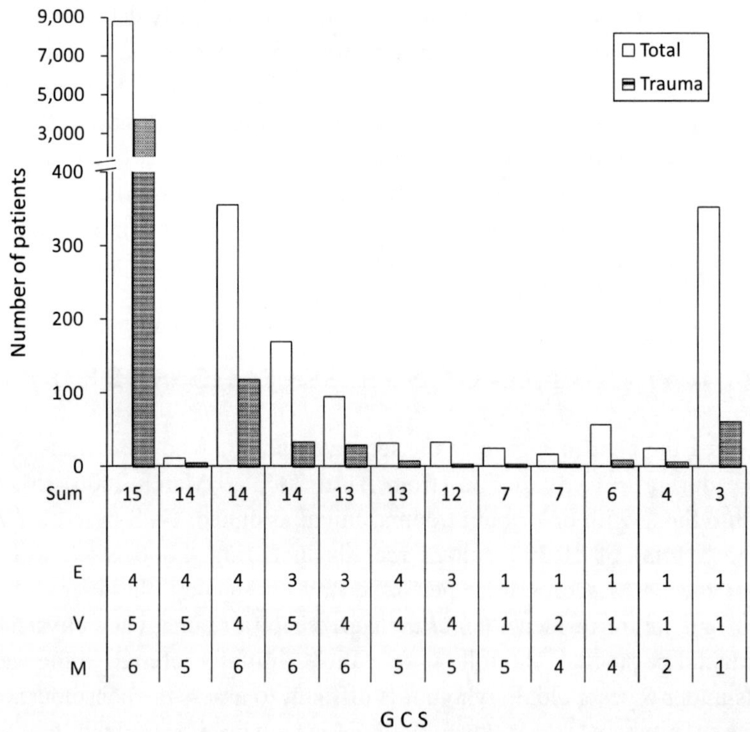

Figure 2. The most common combinations of GCS eye (E), verbal (V), and motor (M) responses, representing initial consciousness level in ED patients. The top 10 EVM combinations (E3V4M5, E1V1M5, and E1V2M4 ranked 10th) that were initially assigned to emergency trauma patient are shown.

The total GCS scores of the ED patients derived from the sum of the EVM scores were shown in Figure 1. Concerning the main reason for calling of the ambulance, the ratio of illness and trauma was almost 6:4. A total GCS score of 15 was initially assigned to 80% of illness, 92% of trauma, and 85% of the total cases of ED patients. Emergency trauma patients were assigned 41 EVM combinations in this study out of 120 mathematically possible combinations (4 x 5 x 6 levels of E, V, and M). The top ten EVM combinations assigned to trauma patients are shown in Figure 2. The EVM combination of E4V5M6 (a total GCS score of 15) was the most frequently assigned to both the total cases of ED patients and the emergency trauma patients. The second and the third most common EVM combination was E4V4M6 (a total GCS score of 14) and E1V1M1 (a total GCS score of 3), respectively.

5. GCS Video Simulation Testing

A video simulation was made in which the eight clinically frequent combinations of EVM responses (listed in Figure 3) were represented by mock patients. The number of patients in the ED who were assigned to the selected eight EVM profiles accounted for 98% of all the trauma cases. The roles of mock patients and physicians assessing consciousness levels in the video were played by medical students under the supervision of a board-certified neurosurgeon and an emergency physician. The faculty of the department including three emergency physicians who had experience over ten years reviewed the video performance and agreed that the mock patients adequately played the levels of consciousness and were assessed their consciousness levels according standard examination. The video performance was also verified by good inter-rater reliability tested among trained emergency physicians (answers of three attendings who had completed the course of Advanced Trauma Care were in 100% agreement with the correct EVM scores depicted in the video, and answers of three other attendings were in 87.5% agreement and did not agree with a score of E1V1M4 or E3V5M6 out of eight scenarios). The video clip can be viewed on the *Clinical Neurology and Neurosurgery* website [5].

We performed a prospective observational study using the video simulations. PGY-1 junior residents (n=94) who were headed at the hospital and rotated into the ED during a 21-month period (from June 2006 to February 2008) were subjected to the study as a uniform group of inexperienced physician and tested for GCS score assignment. The medical training system comprises six-year medical school after graduation from high school and two-year clinical internship (PGY-1 and 2; called a junior resident) under the control of the Ministry. Residency programs of specific departments follow PGY-2 (a period depends on a program; called a senior resident).

At the beginning of off-the-job training programs during the PGY-1 ED rotation, the video was shown in a classroom setting to a group consisting of 7 to 10 junior residents. A scoring sheet, on which only the scales of the three components of the GCS (1-4 for “E”, 1-5 for “V”, and 1-6 for “M”) were printed for the eight scenarios, was provided. The trainees were allowed to watch each scenario twice and then filled in their interpretations on the answer sheets. As the interval between each scenario was only a couple of seconds, they had to record their answers while the scenario was repeated. The video took 11 minutes including repeats and intervals. They were provided with no instruction in the GCS prior to watching the pre-test video and not allowed to see any written references during the test. The answer sheets were subsequently collected from them. After the pre-test they were then instructed on the interpretation of GCS scoring on the video simulations that they had seen. Finally, they were post-tested on the GCS assessment of the same eight EVM combinations by means of a different video scenario. The pre- and post-test were carried out only for the educational purpose of effective instruction, and the results of the examination were not reported on the residents' personal records. Therefore, the data obtained from the PGY-1 junior residents in the study are likely to represent the overall quality of GCS assessment by inexperienced physicians.

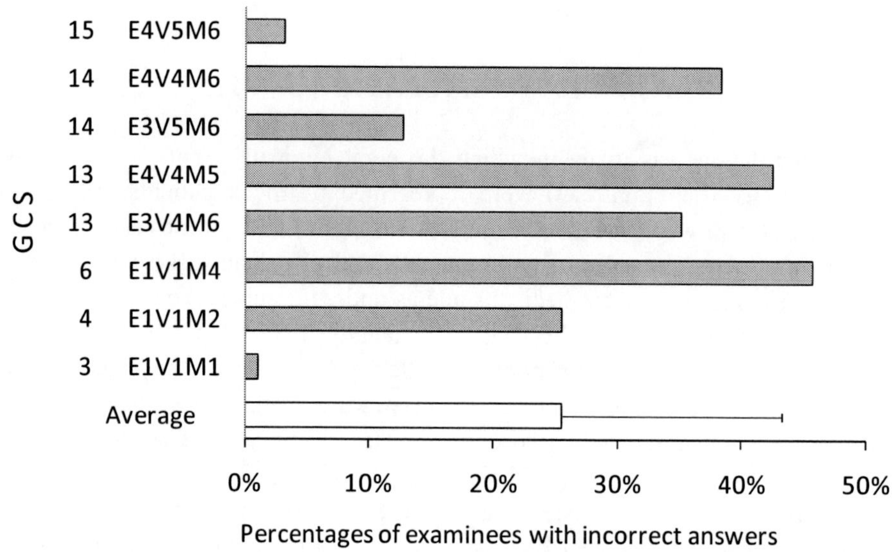

Figure 3. Inaccuracy of GCS scores when assessed by PGY-1 junior residents. Video simulations represented eight clinically frequent combinations of GCS eye (E), verbal (V), and motor (M) responses. PGY-1 junior residents (n=94) took a GCS assessment examination of the eight different consciousness levels. An error bar represents standard deviation.

The primary data of the pre-test answer sheets were collated. One examinee presented the answer sheet without having circled any scores. There were no sheets on which two or more circles were marked on numbers of any scale of the E, V, or M components. Numbers of incorrect answers for each EVM combination were tabulated. The EVM combinations incorrectly assigned by the examinees were also determined. We calculated the difference between the correct composite GCS scores and the incorrect answers, and the extents and frequencies of overestimation and underestimation. To elucidate the misjudged factors of GCS scores, we classified incorrect answers into their individual EVM component scores.

6. Inaccuracy of GCS Scores

For an average of the eight different presentations of the consciousness levels in the video simulation, 26% ± 18% (average ± standard deviation) of the PGY-1 junior residents failed to answer correctly regarding the EVM responses based on the GCS (Figure 3). The examinees seemed to accurately judge consciousness levels of fully conscious (E4V5M6, a total GCS score of 15) and deeply comatose (E1V1M1, a total GCS score of 3) patients. However, the other six combinations of EVM responses were often misjudged. Over 40% of PGY-1 junior residents failed to answer correctly E4V4M5 and E1V1M4. Importantly, 38% of examinees were unable to correctly assess the E4V4M6 level (a total GCS score of 14), which was assigned to a substantial number of ED patients (Figure 2).

7. Incorrect Assignment of GCS Component Profiles; Eye, Verbal, and Motor Scores

There were 192 incorrect answers out of 752 questions total (each of 94 examinees underwent 8 questions). The EVM profile that was incorrectly assigned most often was E4V5M6 (a total GCS score of 15), and 27 out of 30 incorrect assignments of E4V5M6 occurred when assessing a mock patient who was simulating E4V4M6 (Table 2). This suggests that quite a large number of patients with slightly disturbed consciousness had been mistakenly misjudged as normal (GCS 15). This error results in undertriage of minor head trauma patients in clinical practice. In contrast, E1V1M1 was incorrectly assigned only 4 times when assessing mock patients who were simulating E1V1M2 and E1V1M4 (Table 2). The E1V1M1 simulation was only misjudged by the one examinee (Figure 3) that provided the blank answer sheet.

Table 2. Incorrect assignments for GCS eye (E), verbal (V), and motor (M) response scores in the video examination assessing consciousness levels.

EVM scenario	Incorrect answer	Number of incorrect answers*	
E4V4M6		27	
E4V4M5	E4V5M6	1	30 (32%)
E3V4M6		2	
E4V5M6		2	
E3V5M6	E4V4M6	1	5 (5%)
E3V4M6		2	
E3V4M6	E3V5M6	15	15 (16%)
E4V4M6	E3V4M6	2	9 (10%)
E3V5M6		7	
E1V1M2	E1V1M4	2	2 (2%)
E1V1M4	E1V1M2	14	14 (15%)
E1V1M4	E1V1M1	2	4 (4%)
E1V1M2		2	
Any	Other EVM combinations or blank answer	113	113 (-)
Total			192

*The number of incorrect assignments for each EVM combination was divided by 94 (i.e., the number of questions of each EVM combination for the 94 examinees).

8. Overestimation and Underestimation of Total GCS Scores

Because a total GCS score, rather than EVM profiles, is widely used as a rough estimate of the overall neurological status of a patient, we also investigated the discrepancy in a total GCS score resulting from the misjudgment of the EVM profiles. A total of 192 misjudgments consisted of 78 overestimations (10% of the total number of questions), 101 underestimations (13% of the total number of questions) of consciousness level, and 8 blank answers (Figure 4). Clinically, over- and underestimations of a total GCS score correspond to under- and overtraiage of severity of brain injury, respectively. Misjudgments did not change the total GCS score in only 5 cases (less than 1% of the total number of questions), in which one point misjudgment of the score on each of two EVM profile components cancelled each other out (e.g., E4V5M5 versus E4V4M6; the total GCS score is 14 in both cases). Regarding the eight different consciousness levels simulated in the video, 1 point of over- and underestimation on the total GCS score accounted for approximately 10% each in judgments of consciousness level, and discrepancies of 2 or more points accounted for 5% in total.

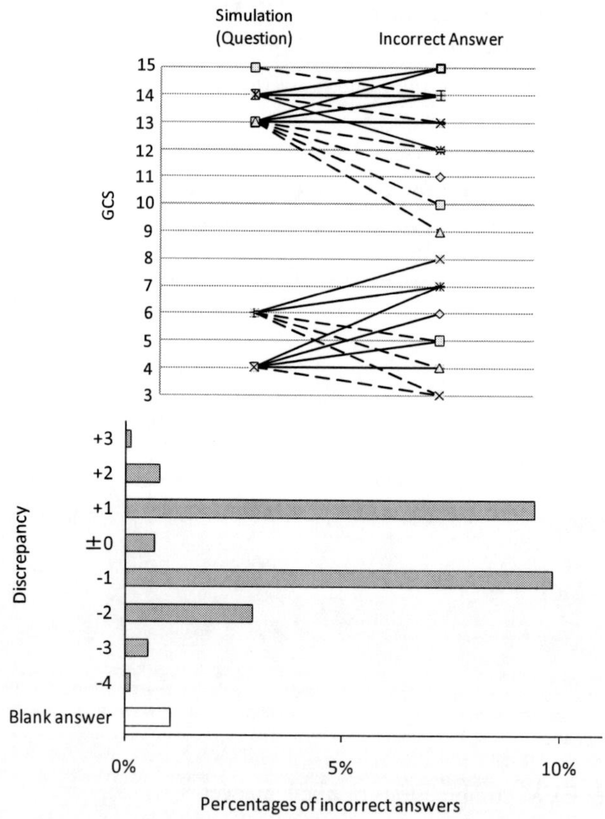

Figure 4. Overestimation and underestimation of total GCS scores. (Upper panel) Total GCS scores of incorrect answers corresponding to questions. (Lower panel) Discrepancy between the correct total GCS scores and the incorrect answers. The number of incorrect answers in the video examination assessing consciousness levels was divided by 752 (i.e., the total number of questions, 8 scenarios x 94 examinees).

9. Misjudged Factors of GCS

We used video simulations and administered a GCS assessment examination to a large number of PGY-1 junior residents representing a uniform group of inexperienced physicians, in order to address the primary factors that contribute to GCS scoring misinterpretation.

Simulation		Incorrect answer	
EVM component and score	EVM combination	EVM component and score	EVM combination
E3	E3V5M6 E3V4M6	E4	E4V5M6 E4V4M6
V5	E4V5M6 E3V5M6	V4	E4V4M6 E3V4M6
V4	E4V4M6 E4V4M5 E3V4M6	V5	E4V5M6 E4V5M5 E3V5M6
V4	E4V4M6 E4V4M5 E3V4M6	V3	E4V3M6 E4V3M5 E3V3M6
M5	E4V4M5	M4	E4V4M4
M4	E1V1M4	M5	E1V1M5
M4	E1V1M4	M3	E1V1M3
M4	E1V1M4	M2	E1V1M2
M2	E1V1M2	M3	E1V1M3

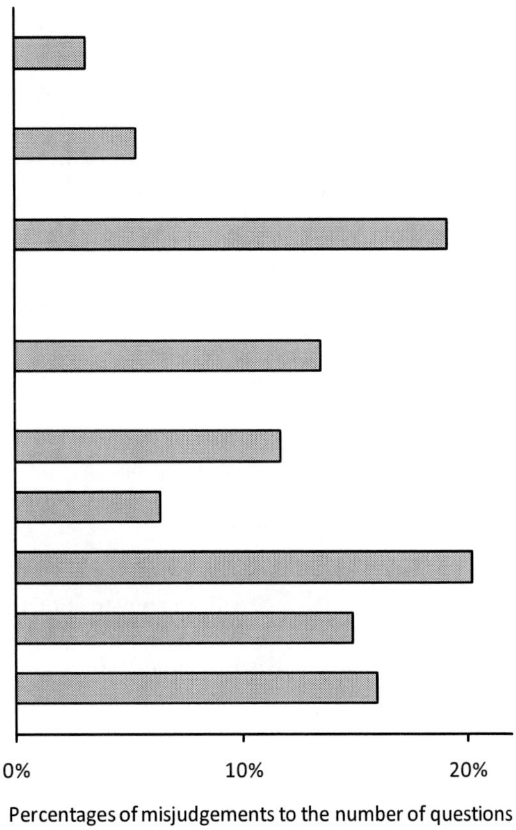

Figure 5. Misjudgments of GCS eye (E), verbal (V), and motor (M) response scores. EVM component scores of simulation that comprised more than five incorrect answers are listed. The total number of incorrect answers was divided by the total number of questions (i.e., number of scenarios x 94 examinees).

Previous GCS scoring studies that have utilized a videotape have been reported [14, 21]. However, these studies were conducted in order to test the interrater reliability of the GCS and recorded actual patient examination in which vague responses might occur. We investigated the individual components of EVM profiles with respect to incorrect answers on the video examination, and were able to identify specific errors in GCS assessment. Incorrect answers were classified into their individual EVM components and scores (Figure 5).

Incorrect scoring of V4 as V5 regarding the best verbal response was the most common error, occurring in 54 out of the 192 incorrect EVM profiles. The V4 performance was represented in three scenarios of E4V4M6, E4V4M5, and E3V4M6 in the video. Thus, 19% of simulations including V4 were misjudged as V5 regarding the verbal GCS score (54 incorrect answers were divided by 282 questions).V4 was also mistaken for V3 in 38 incorrect answers. These misjudgments can be grouped into the assessment of "confused conversation (V4)" as a specific error of GCS scoring, and this category of misjudgment was the most frequent fault in the knowledge of the GCS definition (a total of 92 out of 192 incorrect answers). There were 10 incorrect assignments of V5 as V4, and this scoring error was associated with the assessment of V4 as well.

The assessment of "withdrawal (M4)" regarding the best motor response was the second most frequently misjudged factor. M4 was incorrectly assessed as M5, M3, or M2, comprising 39 out of 192 incorrect answers (Figure 5). Thus, 41% of examinees misjudged the motor score of the E1V1M4 scenario. A scoring error of M5 as M4 was also associated with the assessment of M4. Because we reviewed the database of ED patients and selected the most common eight GCS presentation, the M4 performance was represented only in one scenario of E1V1M4. If the distribution of the consciousness levels of ED patients is not considered, the most frequent misjudged factor was M4.

Table 3. Glasgow Coma Scale with explanations [5].

Component	Level	Score
Eye opening	Spontaneous	E4
	To speech (then reverts to previous state after cessation of stimulation)	3
	To pain	2
	None	1
Best verbal response	Oriented (be able to recognize time [date], place, and person)	V5
	Confused conversation (orientation-impaired, but can speak a sentence consisting of more than a couple of words)	4
	Inappropriate words	3
	Incomprehensible sounds	2
	None	1
Best motor response	Obeys commands	M6
	Localizes pain (patient's arm moves to the point of painful stimuli on the sternum or the supraorbital notch etc.)	5
	Withdrawal (normal flexion; patient's axilla opens following painful stimuli on the nail bed or to the arm.)	4
	Abnormal flexion (decorticate; patient's elbows, wrists and fingers are flexed with closed axillae.)	3
	Extension (decerebrate; patient's arms are extended by the sides with extended elbows.)	2
	None	1

10. Conclusions

The major errors in GCS score assignment were grouped into two categories: the assessment of "confused conversation (V4)" regarding the best verbal response, and the assessment of "withdrawal (M4)" regarding the best motor response. Additional instruction regarding the specific misjudged factors described in this chapter may help inexperienced medical personnel improve the reliability of GCS score assignment to casualties or patients with consciousness disturbance. From this educational point of view, we developed an example of the GCS with additional descriptions as shown in Table 3 [5].

References

[1] Teasdale, G. and B. Jennett, Assessment Of Coma And Impaired Consciousness. A Practical Scale. *Lancet*, 1974. 2(7872): P. 81-4.

[2] Jennett, B. and G. Teasdale, Aspects Of Coma After Severe Head Injury. *Lancet*, 1977. 1(8017): P. 878-81.

[3] Teasdale, G., et al., Adding Up The Glasgow Coma Score. *Acta Neurochir Suppl* (Wien), 1979. 28(1): P. 13-6.

[4] Teoh, L.S., et al., Glasgow Coma Scale: Variation In Mortality Among Permutations Of Specific Total Scores. *Intensive Care Med,* 2000. 26(2): P. 157-61.

[5] Namiki, J., et al., Inaccuracy And Misjudged Factors Of Glasgow Coma Scale Scores When Assessed By Inexperienced Physicians. *Clin Neurol Neurosurg,* 2011. 113(5): P. 393-8.

[6] NICE Clinical Guideline Development Group, Selecting Patients For CT Imaging Of The Head., In NICE Clinical Guideline 56 Head Injury: Triage, Assessment, Investigation And Early Management Of Head Injury In Infants, Children And Adults, N.C.C.F.A. Care, Editor. 2007, *National Institute For Health And Clinical Excellence: London*. P. 22-4.

[7] Dunning, J., et al., A Meta-Analysis Of Clinical Correlates That Predict Significant Intracranial Injury In Adults With Minor Head Trauma. *J Neurotrauma*, 2004. 21(7): P. 877-85.

[8] Fabbri, A., et al., Observational Approach To Subjects With Mild-To-Moderate Head Injury And Initial Non-Neurosurgical Lesions. *J Neurol Neurosurg Psychiatry,* 2008. 79(10): P. 1180-5.

[9] NICE Clinical Guideline Development Group, Community Health Services And NHS Minor Injury Clinics In NICE Clinical Guideline 56 Head Injury: Triage, Assessment, Investigation And Early Management Of Head Injury In Infants, Children And Adults, N.C.C.F.A. Care, Editor. 2007, *National Institute For Health And Clinical Excellence: London.* P. 14-6.

[10] Crossman, J., et al., The Glasgow Coma Score: Reliable Evidence? *Injury*, 1998. 29(6): P. 435-7.

[11] Lindsay, K.W., G.M. Teasdale, and R.P. Knill-Jones, Observer Variability In Assessing The Clinical Features Of Subarachnoid Hemorrhage. *J Neurosurg,* 1983. 58(1): P. 57-62.

[12] Rowley, G. and K. Fielding, Reliability And Accuracy Of The Glasgow Coma Scale With Experienced And Inexperienced Users. *Lancet*, 1991. 337(8740): P. 535-8.

[13] Teasdale, G., R. Knill-Jones, and J. Van Der Sande, Observer Variability In Assessing Impaired Consciousness And Coma. *J Neurol Neurosurg Psychiatry,* 1978. 41(7): P. 603-10.

[14] Juarez, V.J. and M. Lyons, Interrater Reliability Of The Glasgow Coma Scale. *J Neurosci Nurs,* 1995. 27(5): P. 283-6.

[15] Tesseris, J., et al., A Comparative Study Of The Reaction Level Scale (RLS85) With Glasgow Coma Scale (GCS) And Edinburgh-2 Coma Scale (Modified) (E2CS(M)). *Acta Neurochir* (Wien), 1991. 110(1-2): P. 65-76.

[16] Elliott, M., Interrater Reliability Of The Glasgow Coma Scale. *J Neurosci Nurs,* 1996. 28(4): P. 213-4.

[17] Gill, M., et al., Interrater Reliability Of 3 Simplified Neurologic Scales Applied To Adults Presenting To The Emergency Department With Altered Levels Of Consciousness. *Ann Emerg Med*, 2007. 49(4): P. 403-7.

[18] Gill, M.R., D.G. Reiley, and S.M. Green, Interrater Reliability Of Glasgow Coma Scale Scores In The Emergency Department. *Ann Emerg Med*, 2004. 43(2): P. 215-23.

[19] Heard, K. and V.S. Bebarta, Reliability Of The Glasgow Coma Scale For The Emergency Department Evaluation Of Poisoned Patients. *Hum Exp Toxicol,* 2004. 23(4): P. 197-200.

[20] Holdgate, A., N. Ching, and L. Angonese, Variability In Agreement Between Physicians And Nurses When Measuring The Glasgow Coma Scale In The Emergency Department Limits Its Clinical Usefulness. *Emerg Med Australas*, 2006. 18(4): P. 379-84.

[21] Menegazzi, J.J., et al., Reliability Of The Glasgow Coma Scale When Used By Emergency Physicians And Paramedics. *J Trauma,* 1993. 34(1): P. 46-8.

[22] Brain Trauma Foundation And American Association Of Neurological Surgeons, Early Indicators Of Prognosis In Severe Traumatic Brain Injury. Glasgow Coma Scale Score. *J Neurotrauma,* 2000. 17(6-7): P. 563-71.

[23] Riechers, R.G., 2nd, et al., Physician Knowledge Of The Glasgow Coma Scale. *J Neurotrauma,* 2005. 22(11): P. 1327-34.

[24] Marion, D.W. and P.M. Carlier, Problems With Initial Glasgow Coma Scale Assessment Caused By Prehospital Treatment Of Patients With Head Injuries: Results Of A National Survey. *J Trauma*, 1994. 36(1): P. 89-95.

[25] Meredith, W., et al., The Conundrum Of The Glasgow Coma Scale In Intubated Patients: A Linear Regression Prediction Of The Glasgow Verbal Score From The Glasgow Eye And Motor Scores. *J Trauma,* 1998. 44(5): P. 839-44; Discussion 844-5.

[26] Rutledge, R., et al., Appropriate Use Of The Glasgow Coma Scale In Intubated Patients: A Linear Regression Prediction Of The Glasgow Verbal Score From The Glasgow Eye And Motor Scores. *J Trauma,* 1996. 41(3): P. 514-22.

[27] Buechler, C.M., et al., Variation Among Trauma Centers' Calculation Of Glasgow Coma Scale Score: Results Of A National Survey. *J Trauma*, 1998. 45(3): P. 429-32.

[28] Sternbach, G.L., The Glasgow Coma Scale. *J Emerg Med*, 2000. 19(1): P. 67-71.

In: Comas and Syncope: Causes, Prevention and Treatment ISBN 978-1-62100-603-9
Editors: Eduardo Silva and Gabriel Cruz.

Chapter 9

CLASSIFICATION OF SYNCOPE-PRODUCING CARDIAC ARRHYTHMIAS

Gaetano Santulli
Columbia University Medical Center
College of Physicians & Surgeons
New York, NY, US

ABSTRACT

Syncope is ***a clinical syndrome that is a subset of a broader range of conditions that cause*** sudden and temporary loss of consciousness, from which recovery is spontaneous.

The causes of syncope can be grouped into four major categories: neurologic, metabolic, vasomotor and cardiac. Of these, only cardiac syncope commonly leads to sudden death.

There are two major groups of cardiac syncope: obstructive cardiac lesions and cardiac arrhythmias. Cardiac arrhythmias represent most common cause and can be subdivided in bradycardias and tachycardias. The aim of this article is to provide a clear overview of syncope-producing cardiac arrhythmias.

INTRODUCTION

Syncope is ***defined as*** a transient loss of consciousness due to transient global cerebral hypoperfusion characterized by rapid onset, short duration, and spontaneous complete recovery [1].

It is common - one out of every 4 persons experiences a faint during his or her lifetime - and often does not indicate a serious medical problem. However, sometimes syncope indicates a dangerous or even life-threatening condition, so when syncope occurs it is important to figure out the cause. And the first order of business is to make sure the patient presenting with syncope isn't about to die [2]. The term syncope derives from the Greek word synkoptein, meaning "to cut short," and purportedly, Hippocrates himself provided the first description of a patient with the disorder [3].

It is estimated that as many as 6% of hospitalizations and 3% of emergency visits are prompted by transient loss of consciousness. In additional, 75% of patients with recurring syncope will alter their daily lives, nearly 40% will change jobs, and 75% will suffer from depression or anxiety, especially if not diagnosed and treated. [4-6].

Although individual differences in susceptibility to syncope exist, almost any healthy person can faint given the proper circumstances. The fact that simple faints are so common probably reflects incomplete evolutionary adaptation of humans to the upright body position. Giraffes have done an excellent job of mastering gravity and can raise their head from the ground to an elevation of 18 feet in one second without passing out [7]. Standing upright requires special blood pressure regulation to drive blood against gravity up to the head. Most patients experience warning symptoms telling them they are about to pass out (pre-syncope) [8-11]. Premonitory symptoms such as palpitations, weakness, confusion, and nausea allow an individual to prepare for collapse without bodily harm. However, elderly patients may be prone to serious injury such as hip fractures [12, 13].

Widely used definitions of syncope are unhelpful. For example, in many medical textbooks, and in the Framingham study, syncope is defined as "a sudden loss of consciousness associated with the inability to maintain postural tone, followed by spontaneous recovery"[14]. This definition of syncope is so wide that it encompasses almost all causes of transient unconsciousness, including concussion and several types of epileptic seizures, transient ischemic attacks and stroke [15-18].

A more precise definition was recently provided by the Task Force on Syncope of the European Society of Cardiology (ESC): "***Syncope*** is a transient loss of consciousness due to transient global cerebral hypoperfusion characterized by rapid onset, short duration, and spontaneous complete recovery"[1]. The term global does not necessarily imply that the loss of consciousness is due to dysfunction of the entire brain; this may be due to brain stem impairment, dysfunction of a major part of both hemispheres, or both. This definition of syncope differs from others by including the cause of unconsciousness, i.e. transient global cerebral hypoperfusion. Without that addition, the definition of syncope becomes wide enough to include disorders such as epileptic seizures and concussion [19].

The episodic and unpredictable nature of syncope makes its study and differential diagnosis difficult. Because syncope includes rather disparate disease process, the diagnostic evaluation requires specialized knowledge in different fields including cardiac electrophysiology, blood pressure and blood flow regulation (hemodynamics, biology of blood vessels), hormone regulation (endocrinology), and integrative neurophysiology (peripheral and central nervous system, brain research). Therefore the diagnosis and management of syncope demand teamwork by specialists. However, a prudent approach is for the patient to consult a clinical cardiologist who has the knowledge and tools to rule out life-threatening arrhythmias [1]. Numerous conditions associated with loss of consciousness are often thought to be part of differential diagnosis, but are not true syncope. Disease processes characterized by sustained low blood pressure (hypotensive states or shock syndromes) include aortic dissection (aortic rupture), shock due to heart attacks or failure (cardiogenic shock), cardiac tamponade (compression of the heart by acute fluid accumulation in the pericardial cavity), massive pulmonary embolism, and others are not associated with rapid spontaneous recovery. In these syndromes, impairment of consciousness and reduction in brain blood flow are sustained. Brain function and consciousness may be perturbed when the blood contains low concentrations of oxygen (hypoxemia), glucose (hypoglycemia), calcium (hypocalcemic tetany), or hydrogen ions (alkalotic tetany). An excess of hydrogen ions (acidosis, hypercapnia) may also affect consciousness. With some exceptions, such as hypoglycemia or low carbon dioxide due to voluntary hyperventilation, loss of consciousness associated with changes in blood chemistry are neither transient nor acutely reversible. When the blood composition is therapeutically corrected, one characteristically observes gradual recovery in brain function over minutes to hours [2, 11, 19].

Consciousness depends upon the normal activity of the lower brain (brain stem). Therefore, transient loss of consciousness independent of changes in brain blood flow or chemistry may be caused by disturbances in the activity of brain cells themselves (seizure disorders). As a rule, one should never ascribe spells with syncope-like features to a rare seizure disorder until a cardiologist rules out a cardiac arrhythmia [19, 20].

The Role of the Heart

The causes of syncope can be grouped into four major categories: neurologic, metabolic, vasomotor and cardiac [21]. ***However, despite frequent confusion in the medical literature, syncope should be clearly distinguished from those other causes of transient loss of counsciousness by virtue of its pathophysiology; specifically, syncope is the result of self-terminating inadequacy of global cerebral nutrient perfusion, while the other forms have different etiologies [22].***

In other words, albeit a number of disease processes can cause syncope, hypoperfusion of the cerebral cortexes and reticular activating system is the common final pathway for the majority of episodes [23]. Interruption of cerebral blood flow for 8 to 10 seconds usually produces loss of consciousness. In contrast, seizures, which are difficult to distinguish from syncope, impair consciousness by disturbing cerebral electrical activity without necessarily affecting perfusion [19, 24].

Only cardiac syncope commonly leads to sudden death [1]. There are two major varieties of cardiac syncope: obstructive cardiac lesions, and cardiac arrhythmias [2].

As regards obstructive lesions, several heart disorders can result in an obstruction of blood flow through the heart. These include obstructed heart valves (aortic stenosis and mitral stenosis are the most common examples; obstructed blood vessels (such as a massive pulmonary embolus; and cardiac tumors (such as an atrial myxoma, a benign tumor that can obstruct the mitral valve).

Most of these lesions are readily apparent to a physician taking a careful medical history and doing a careful cardiac examination. For the most part, they are readily confirmed by performing a simple echocardiogram [25].

On the other hand, cardiac arrhythmias represent most common cause of cardiac syncope. There are two major groups of life-threatening, syncope-producing cardiac arrhythmias: the bradycardias and the tachycardias.

BRADYCARDIAS

Bradycardia (from the greek: βραδυκαρδία, "heart slowness") is the resting heart rate of under 60 beats per minute, though it is seldom symptomatic until the rate drops below 50 beats per minute. Of interest, trained athletes or young healthy individuals may also have a slow resting heart rate (e.g. professional cyclist Miguel Indurain had a resting heart rate of 28 beats per minute) [26, 27]. The specialized conduction system is responsible for transmitting the heart's electrical impulses from the atria to the ventricles. Disease in the AV node, bundle of His, or the bundle branches can lead to a condition called "heart block." Heart block occurs when the electrical impulses in the atria are stopped from reaching the ventricles. The heart rate can reach dangerously low levels when heart block is present [28].

Bradycardia can be atrial, atrioventricular or ventricular.

The most common atrial bradycardia is sinus bradycardia, usually found in young and healthy adults, in which the symptoms represent with the individual's respirations. On the other hand, a relatively uncommon atrial bradycardia is *Sick sinus syndrome*, also called sinus node dysfunction. It is a group of arrhythmias presumably caused by a malfunction of the sinus atrial node, the heart's primary pacemaker, more frequent in elderly adults, where the cause is often a non-specific, scar-like degeneration of the cardiac conduction system, resulting in many abnormal heart rhythms, including sinus arrest and sinus node exit block. Most sinus node disease is related to simple deterioration in sinus node function caused by aging [29]. Sick sinus syndrome is the most common indication for implanting a permanent pacemaker.

Atrioventricular bradycardia or junction rhythm is usually caused by the absence of the electrical impulse from the sinus node. This usually appears on an ECG with a normal QRS complex accompanied with an inverted P wave either before, during, or after the QRS complex. This is a protective mechanism for the heart, to compensate for a sinus atrial node that is no longer handling the pacemaking activity, and is one of a series of backup sites that can take over pacemaker function when the SA node fails to do so. A junctional escape complex could be also a normal response that may result from excessive vagal tone on the sinus node [30].

A ventricular bradycardia, also known as ventricular escape rhythm or idioventricular rhythm, is a heart rate of less than 50 beats a minute. Also this is a safety mechanism that arises when there is lack of electrical impulse or stimuli from the atrium. Impulses originating from or below the His bundle, also known as ventricular, will produce a wide QRS complex with heart rates between 20 and 40 beats a minute. Those above the His bundle, also known as junctional (see above), will typically range between 40 and 60 bpm with a narrow QRS complex. In a third degree heart block, approximately 61% take place at the bundle branch-Purkinje system, 21% at the atrioventricular node, and 15% at the His bundle [31].

Bradycardia can be underlain by several causes, not only cardiac (ischemic heart disease, vascular heart disease, valvular heart disease, or degenerative primary electrical disease), which can involve drug use or abuse, metabolic or endocrine issues, especially in the thyroid, autoimmunity, electrolyte imbalance, neurologic factors, autonomic reflexes and situational factors. Ultimately, the causes act by three mechanisms: depressed automaticity of the heart, conduction block, or escape pacemakers and rhythms.

Tachycardias

Tachycardia (from the greek: ***Ταχυκαρδία***, "heart rapidness") typically refers to a heart rate that exceeds the normal range for a resting heart rate (in adults, more than 100 beats per minute). When the heart beats excessively rapidly, the heart pumps less efficiently and provides less blood flow to the rest of the body, including the heart itself [32, 33]. The increased heart rate also leads to increased work and oxygen demand by the heart, which can lead to rate related ischemia.

An ECG is commonly used to classify the type of tachycardia. They may be classified into SupraVentricular and Ventricular. The first rarely causes syncope, except in Wolff-Parkinson-White (WPW) syndrome.

WPW syndrome is a type of atrioventricular reentrant tachycardia due to an accessory pathway known as the bundle of Kent that leads to a pre-excitation of ventricles. This accessory pathway is an abnormal electrical communication from the atria to the ventricles. The incidence of WPW syndrome is between 0.1% and 0.3% of the general population [34]. While the majority of individuals with a bundle of Kent remain asymptomatic throughout their entire lives, there is a risk of sudden death associated with the syndrome [35].

Another classification is the subdivision of tachycardias into narrow and wide complex based on the QRS complex (the cut-off is 120 msec) [36]. Among tachycardias with a narrow QRS complex there are Sinus tachycardia, Atrial fibrillation, Atrial flutter, Atrial tachycardia, Multifocal atrial tachycardia, AV nodal reentrant tachycardia, Junctional tachycardia.

Sinus tachycardia is a rhythm in which the rate of impulses arising from the SA node is regular but elevated. It is usually a physiological response to several situations, such as exercise, stress, fright, anger, flight [37, 38].

Atrial fibrillation is one of the most common cardiac arrhythmias. It is generally an irregular, narrow complex rhythm. However, it may show wide QRS complexes on the ECG if a bundle branch block is present. At high rates, the QRS complex may also become wide due to the Ashman phenomenon. It may be difficult to determine the rhythm's regularity when the rate exceeds 150 beats per minute. Depending on the patient's health and other variables such as medications taken for rate control, atrial fibrillation may cause heart rates that span from 50 to 250 beats per minute (or even higher if an accessory pathway is present) [39].

Atrial flutter is the most common type of macro-reentrant atrial tachycardia. It was first identified as an independent medical condition in 1920 by the British physician Sir Thomas Lewis. The most common clinicla presentation of patients is palpitation, while syncope is rare. There are two types of atrial flutter, the common type I and rarer type II.

The atrial rate during Type I atrial flutter, also known as common or typical atrial flutter, is usually 250 to 350 beats per minute. The reentrant loop circles the right atrium, passing through the cavo-tricuspid isthmus (a body of fibrous tissue in the lower atrium between the inferior vena cava and the tricuspid valve). Type I flutter is further divided into two subtypes, known as counterclockwise atrial flutter and clockwise atrial flutter depending on the direction of current passing through the loop.

Type II flutter follows a significantly different re-entry pathway to type I flutter, and is typically faster, usually 340-350 beats per minute. Left atrial flutter is common after incomplete left atrial ablation procedures [40, 41].

Atrial tachycardia is defined as a supraventricular tachycardia that does not require the AV junction, accessory pathways, or ventricular tissue for initiation and maintenance of the tachycardia. The atrial rhythm is usually regular. The conducted ventricular rhythm is also usually regular but may become irregular, often at higher atrial rates because of variable conduction through the AV node, thus producing different conduction patterns.

Multifocal atrial tachycardia (MAT) is an arrhythmia with an irregular atrial rate greater than 100 bpm. Atrial activity is well organized, with at least 3 morphologically distinct P waves, irregular P-P intervals, and an isoelectric baseline between the P waves. MAT has previously been also described by terms such as chaotic atrial rhythm or tachycardia, chaotic atrial mechanism, and repetitive paroxysmal MAT [42-44].

Atrioventricular nodal reentrant tachycardia (AVNRT) is a common cause of palpitations in women. It occurs when a reentry circuit forms entirely within or just next to the AV node. The circuit usually involves two anatomical pathways: the fast pathway and the slow pathway, which are both in the right atrium (they should not be confused with the accessory pathways that give rise to the WPW syndrome). The slow pathway (which is usually targeted for ablation) is located inferiorly and slightly posterior to the AV node, often following the anterior margin of the coronary sinus. The fast pathway is usually located just superior and posterior to the AV node. Patients with AV nodal reentrant tachycardia have no greater incidence of heart disease than the general population [45-48].

Junctional tachycardia is an automatic tachycardia originating in the AV junction. It tends to be regular. The junction tachycardias are rare but still occasionally seen. Two forms of junctional tachycardia are recognized (permanent and automatic), and each has a different mechanism [49]. The permanent form is a reentrant tachycardia in which an accessory pathway lies very close to the AV node. The tachycardia is slower than AVNRT and is often incessant. This tachycardia has been associated with the development of a reversible cardiomyopathy that resolves when the tachycardia is successfully terminated. Automatic junctional tachycardia is, as the name suggests, an automatic tachycardia of the AV node. It may be caused by drug toxicity (a sign of digitalis toxicity) or it may be idiopathic [50].

Wide QRS complex tachycardia can be either the ventricular tachycardia (VT) or supraventricular tachycardia with abnormal conduction to the ventricles. Supraventricular tachycardia with a pre-existing bundle branch block is another form of wide QRS tachycardia. Moreover, when the myocardium is grossly damaged the QRS becomes wide even in a supraventricular rhythm. This can also occur in electrolyte abnormalities like severe hyperkalemia in which the QRS widens and merges with the T wave to produce a 'sine wave pattern'. Certain drugs and toxins can also produce widening of QRS in a supraventricular rhythm. There is no single criterion or combination of criteria that provides complete diagnostic accuracy in evaluating a wide complex tachycardia. Algorithms that perform well in initial reports, with selected populations and experienced ECG analysts, are likely to be less accurate in general practice [51-54].

VT, which involves any tachycardia which originates in the ventricles. This arrhythmia, which originate in the ventricles, causes syncope and can result in sudden death. Ventricular tachycardia, which describes a heart rate of over 100 beats per minute with at least three irregular heartbeats as a sequence of consecutive premature beats, can degenerate into ventricular fibrillation, which requires direct current cardioversion [55].

Arrhythmogenic Right Ventricular Dysplasia/Cardiomyopathy (ARVD/C)

A particular mention in this chapter is due to the relation between Syncope and ARVD/C. This is a a type of nonischemic cardiomyopathy that involves primarily the right ventricle, characterized by hypokinetic areas involving the free wall of the right ventricle, with fibrofatty replacement of the myocardium. The anatomic damage present in ARVD/C modifies cardiac electrical activation and repolarization, with associated arrhythmias originating in the right ventricle.

Almost 30 % cases of ARVD/C occur in families. This may be an underestimation as the genes responsible for this disease may or may not express themselves in affected patients, termed low penetrance. If these genes do express themselves, it may be with varying degrees, a condition known as variable expression. Low penetrance and variable expression make it difficult to trace the disease along a family line. Regardless of the mode of inheritance, it appears that the mutant genes code for proteins that make up desmosomes, which are intracellular adhesion complexes that provide mechanical connections between cardiac myocytes (heart cells). When placed under mechanical stress, the impaired desmosomes cause myocytes to detach from each other leading to cell death. This cell death causes inflammation with scar formation and fat deposition [56, 57].

While the initiating factor of the ventricular arrhythmias is unclear, it may be due to triggered activity or reentry. Ventricular arrhythmias are usually exercise-related, suggesting that they are sensitive to catecholamines. The ventricular beats typically have a right axis deviation. Multiple morphologies of ventricular tachycardia may be present in the same individual, suggesting multiple arrhythmogenic foci or pathways. ARVD/C should be considered in cases of sudden cardiac arrest, occurring particularly during exercise or in subjects (in the absence of overt heart disease) who present with VT arising from the right ventricle, showing a left bundle branch block (LBBB) morphology with an inferior axis [58-62].

CONCLUSION

In the evaluation of syncope, the presence of a structural heart disease (coronary artery disease, congestive heart failure, valvular heart disease, or congenital heart disease) has emerged as the most important factor for predicting the risk of death, as well as the likelihood of arrhythmias. ***Unfortunately, the diagnostic evaluation of syncope remains for the most part poorly managed and inefficient [63, 64]. There is no widely adopted definition or classification of syncope and related disorders. This lack of uniformity harms patient care, research, and medical education [65].***

Careful history-taking, physical examination, and judicious use of tests can not only increase the diagnostic yield of the evaluation, but also increase patient safety and reduced overall healthcare costs [66]. Patients with heart disease or an abnormal electrocardiogram have an increased risk of death within one year, and most arrhythmias are found in these patients [67]. Although the cardiac causes of syncope were shown in to be associated with increased mortality and an increased risk of sudden death, recent analyses [68] show that underlying heart disease, irrespective of the cause of syncope, is the factor associated with an increased risk of death [69].

References

[1] Moya, A., et al., Guidelines for the diagnosis and management of syncope (version 2009). *Eur. Heart J.*, 2009. 30(21): p. 2631-71.

[2] Kapoor, W.N., Syncope. *N. Engl. J. Med.*, 2000. 343(25): p. 1856-62.

[3] Maisel, W.H. and W.G. Stevenson, Syncope--getting to the heart of the matter. *N. Engl. J. Med.*, 2002. 347(12): p. 931-3.

[4] Del Greco, M., et al., Diagnostic pathway of syncope and analysis of the impact of guidelines in a district general hospital. The ECSIT study (epidemiology and costs of syncope in Trento). *Ital. Heart J.,* 2003. 4(2): p. 99-106.

[5] Getchell, W.S., et al., Epidemiology of syncope in hospitalized patients. *J. Gen. Intern. Med.*, 1999. 14(11): p. 677-87.

[6] Perennes, A., et al., Epidemiology, clinical features, and follow-up of patients with syncope and a positive adenosine triphosphate test result. *J. Am. Coll Cardiol.*, 2006. 47(3): p. 594-7.

[7] Badlangana, N.L., et al., Observations on the giraffe central nervous system related to the corticospinal tract, motor cortex and spinal cord: what difference does a long neck make? *Neuroscience*, 2007. 148(2): p. 522-34.

[8] Allcock, L., R. Mulcahy, and D. O'Shea, Recurrent syncope and presyncope. *Lancet*, 2001. 357(9270): p. 1801-2.

[9] Fogel, R.I., J.J. Evans, and E.N. Prystowsky, Utility and cost of event recorders in the diagnosis of palpitations, presyncope, and syncope. *Am. J. Cardiol*, 1997. 79(2): p. 207-8.

[10] Kou, W.H., et al., Immediate reproducibility of tilt-table test results in elderly patients referred for evaluation of syncope or presyncope. *Am. J. Cardiol.*, 1997. 80(11): p. 1492-4.

[11] Mathias, C.J., K. Deguchi, and I. Schatz, Observations on recurrent syncope and presyncope in 641 patients. *Lancet*, 2001. 357(9253): p. 348-53.

[12] Galizia, G., et al., Role of early symptoms in assessment of syncope in elderly people: results from the Italian group for the study of syncope in the elderly. *J. Am. Geriatr. Soc.*, 2009. 57(1): p. 18-23.

[13] Roussanov, O., et al., Outcomes of unexplained syncope in the elderly. *Am. J. Geriatr. Cardiol.*, 2007. 16(4): p. 249-54.

[14] Soteriades, E.S., et al., Incidence and prognosis of syncope. *N. Engl. J. Med.*, 2002. 347(12): p. 878-85.

[15] Carey, B.J., J. de Caestecker, and R.B. Panerai, More on deglutition syncope. *N. Engl. J. Med.*, 1999. 341(17): p. 1316-7.

[16] Schima, W., F. Sterz, and P. Pokieser, Syncope after eating. *N. Engl. J. Med.*, 1993. 328(21): p. 1572.

[17] Piwinski, S.E., Neurocardiogenic syncope. *N. Engl. J. Med.*, 1993. 329(26): p. 1966-7.

[18] Papa, L.J., Treatment of neurocardiogenic syncope. *N. Engl. J. Med.*, 1993. 329(13): p. 970.

[19] McKeon, A., C. Vaughan, and N. Delanty, Seizure versus syncope. *Lancet Neurol.*, 2006. 5(2): p. 171-80.

[20] D'Ascia, C, et al., Cardiac resynchronization therapy response predicts occurrence of atrial fibrillation. *Int. J. Cardiol.*, 2011. 147: p. S37.

[21] Mehta, N., M. Zildany Pinheiro Tavora, and C.A. Morillo, Explaining the Unexplained Causes of Syncope Are We There Yet? *J. Am. Coll Cardiol.*, 2011.

[22] Sheldon, R.S., et al., Standardized approaches to the investigation of syncope: Canadian Cardiovascular Society position paper. *Can. J. Cardiol.*, 2011. 27(2): p. 246-53.

[23] Grubb, B.P., Clinical practice. Neurocardiogenic syncope. *N. Engl. J. Med.*, 2005. 352(10): p. 1004-10.

[24] Bibb, M.H., Neurocardiogenic syncope. *N. Engl. J. Med.*, 1998. 339(25): p. 1857.

[25] D'Ascia, S.L., et al., Cardiac resynchronization therapy response predicts occurrence of atrial fibrillation in non-ischaemic dilated cardiomyopathy. *Europace*, 2011. 13 Suppl. 1: p. i55.

[26] Bassett, D.R., Jr., et al., Comparing cycling world hour records, 1967-1996: modeling with empirical data. *Med. Sci. Sports Exerc.*, 1999. 31(11): p. 1665-76.

[27] Indurain, M., Miguel Indurain: the allergic extraterrestrial. Interview by IJ Ansoteguiand MT Audicana. *Allergy*, 1997. 52(12): p. 3-4.

[28] D'Ascia, C, et al., In patients affected by nonischemic dilated cardiomyopathy, cardiac resynchronization therapy response predicts occurrence of atrial fibrillation. *Cardiovasc Revascul Med.*, 2011. 12: p. e41.

[29] Cooke, J., et al., The changing face of orthostatic and neurocardiogenic syncope with age. *QJM*, 2011.

[30] Fowler, N.O., J.C. Fenton, and G.F. Conway, Syncope and cerebral dysfunction caused by bradycardia without atrioventricular block. *Am. Heart J.*, 1970. 80(3): p. 303-12.

[31] Ector, H., et al., Bradycardia, ventricular pauses, syncope, and sports. *Lancet*, 1984. 2(8403): p. 591-4.

[32] Konstantino, Y., et al., Self-terminated ventricular fibrillation and recurrent syncope. *Am. J. Cardiol.*, 2011. 107(4): p. 638-40.

[33] Santulli, G., Coffee and cardiovascular system. *Cardiology Science*, 2006. 4: p. 186-188.

[34] Mizumaki, K., et al., Early repolarization in Wolff-Parkinson-White syndrome: prevalence and clinical significance. *Europace*, 2011.

[35] Pass, R.H. and S.R. Ceresnak, Wolff-Parkinson-White syndrome and isoproterenol testing in children: a valid adjunct to predict risk? *Circ. Arrhythm. Electrophysiol.*, 2011. 4(1): p. 8-10.

[36] D'Ascia, S.L., et al., Advanced algorithms can lead to electrocardiographic misinterpretations. *Int. J. Cardiol.*, 2010. 141(2): p. e34-6.

[37] Monnig, G., et al., Recurrent syncope triggered by inappropriate sinus tachycardia. *Pacing Clin. Electrophysiol.*, 2004. 27(9): p. 1324-6.

[38] Gatzoulis, K.A., et al., Polymorphic ventricular tachycardia induced during tilt table testing in a patient with syncope and probable dysfunction of the sinus node. *Pacing Clin. Electrophysiol.*, 1995. 18(5 Pt 1): p. 1075-9.

[39] Brignole, M., et al., Role of autonomic reflexes in syncope associated with paroxysmal atrial fibrillation. *J. Am. Coll Cardiol.*, 1993. 22(4): p. 1123-9.

[40] Turitto, G., et al., Atrial flutter with spontaneous 1:1 atrioventricular conduction in adults: an uncommon but frequently missed cause for syncope/presyncope. *Pacing Clin. Electrophysiol.*, 2009. 32(1): p. 82-90.

[41] Kawabata, M., et al., Syncope in patients with atrial flutter during treatment with class Ic antiarrhythmic drugs. *J. Electrocardiol.*, 2001. 34(1): p. 65-72.

[42] Ueng, K.C., et al., Radiofrequency catheter modification of atrioventricular junction in patients with COPD and medically refractory multifocal atrial tachycardia. *Chest*, 2000. 117(1): p. 52-9.

[43] Kastor, J.A., Multifocal atrial tachycardia. *N. Engl. J. Med.*, 1990. 322(24): p. 1713-7.

[44] Shine, K.I., J.A. Kastor, and P.M. Yurchak, Multifocal atrial tachycardia. Clinical and electrocardiographic features in 32 patients. *N. Engl. J. Med.*, 1968. 279(7): p. 344-9.

[45] Yamada, T., et al., A case of atrioventricular nodal reentrant tachycardia: what is the mechanism? *Pacing Clin. Electrophysiol.*, 2009. 32(8): p. 1056-9.

[46] Tanaka, Y., et al., Incidence and mechanism of dislocated fast pathway in various forms of atrioventricular nodal reentrant tachycardia. *Circ. J.*, 2007. 71(7): p. 1099-106.

[47] Rostock, T. and S. Willems, Atrioventricular nodal reentrant tachycardia in the elderly: efficacy and safety of radiofrequency catheter ablation. *Pacing Clin. Electrophysiol.*, 2007. 30(6): p. 828-9.

[48] Bae, J.H., K.R. Park, and K.Y. Kim, Atrioventricular nodal reentrant tachycardia combined with sick sinus syndrome. *Int. J. Cardiol.*, 2002. 83(3): p. 259-61.

[49] Brembilla-Perrot, B., et al., Incidence and mechanism of presyncope and/or syncope associated with paroxysmal junctional tachycardia. *Am. J. Cardiol.*, 2001. 88(2): p. 134-8.

[50] Luca, C., Nonparoxysmal AV junctional tachycardia with 2/1 exit block during atrial fibrillation in the absence of digitalis toxicity. The effect of exercise. *Med. Interne*, 1981. 19(3): p. 247-9.

[51] Vereckei, A., et al., Application of a new algorithm in the differential diagnosis of wide QRS complex tachycardia. *Eur. Heart J.*, 2007. 28(5): p. 589-600.

[52] Scheinman, M.M., Can a new algorithm improve diagnostic accuracy in wide QRS complex tachycardia? *Nat. Clin. Pract. Cardiovasc. Med*, 2007. 4(8): p. 414-5.

[53] Dendi, R. and M.E. Josephson, A new algorithm in the differential diagnosis of wide complex tachycardia. *Eur. Heart J.*, 2007. 28(5): p. 525-6.

[54] Gutierrez-Macias, A., et al., Wide-complex tachycardia: diagnostic value of the Brugada algorithm in emergency medicine. *Acad. Emerg. Med.*, 2001. 8(3): p. 300-1.

[55] Sra, J.S., et al., Comparison of cardiac pacing with drug therapy in the treatment of neurocardiogenic (vasovagal) syncope with bradycardia or asystole. *N. Engl. J. Med.*, 1993. 328(15): p. 1085-90.

[56] Ozben, B., et al., Angiotensin-converting enzyme gene polymorphism in arrhythmogenic right ventricular dysplasia: is DD genotype helpful in predicting syncope risk? *J. Renin Angiotensin Aldosterone Syst.*, 2008. 9(4): p. 215-20.

[57] Mutschelknauss, M., et al., Unusual cause of syncope in young girl: arrhythmogenic right ventricular cardiomyopathy. *Clin. Res. Cardiol.* 99(4): p. 243-5.

[58] Peters, S. and D. Selbig, Early repolarization phenomenon in arrhythmogenic right ventricular dysplasia-cardiomyopathy and sudden cardiac arrest due to ventricular fibrillation. *Europace*, 2008. 10(12): p. 1447-9.

[59] Gattenlohner, S., et al., Sudden cardiac death due to acute right heart failure in a patient with arrhythmogenic right ventricular dysplasia and acute myocardial infarction. *Eur. Heart J.*, 2008. 29(17): p. 2063.

[60] Aziz, S., R.F. McMahon, and C.J. Garratt, Images in cardiovascular medicine. Sudden cardiac death in arrhythmogenic right ventricular dysplasia. *Circulation*, 2000. 101(7): p. 825-7.

[61] Tada, H., et al., Arrhythmogenic right ventricular cardiomyopathy underlies syndrome of right bundle branch block, ST-segment elevation, and sudden death. *Am. J. Cardiol.*, 1998. 81(4): p. 519-22.

[62] Martini, B., C. Basso, and G. Thiene, Sudden death in mitral valve prolapse with Holter monitoring-documented ventricular fibrillation: evidence of coexisting arrhythmogenic right ventricular cardiomyopathy. *Int. J. Cardiol.*, 1995. 49(3): p. 274-8.

[63] Linzer, M., et al., Diagnosing syncope. Part 1: Value of history, physical examination, and electrocardiography. Clinical Efficacy Assessment Project of the American College of Physicians. *Ann. Intern. Med.*, 1997. 126(12): p. 989-96.

[64] Lewis, S.L. and M. O'Toole, Familial occurrence of neurocardiogenic syncope. *N. Engl. J. Med.*, 1994. 331(22): p. 1529.

[65] Mitro, P., et al., A prospective study of the standardized diagnostic evaluation of syncope. *Europace*, 2011. 13(4): p. 566-71.

[66] Oh, J.H., B.H. Hanusa, and W.N. Kapoor, Do symptoms predict cardiac arrhythmias and mortality in patients with syncope? *Arch. Intern. Med.*, 1999. 159(4): p. 375-80.

[67] D'Ascia, S.L., et al., Cardiac resynchronization therapy response predicts occurrence of atrial fibrillation in non-ischemic dilated cardiomyopathy. *Int. J. Clin. Pract.*, 2011. In press.

[68] Sarasin, F.P., et al., A risk score to predict arrhythmias in patients with unexplained syncope. *Acad. Emerg. Med.*, 2003. 10(12): p. 1312-7.

[69] Kapoor, W.N., Syncope-past, present and future. *Clin. Auton. Res.*, 2004. 14 Suppl 1: p. 1-3.

In: Comas and Syncope: Causes, Prevention and Treatment ISBN 978-1-62100-603-9
Editors: Eduardo Silva and Gabriel Cruz.

Chapter 10

THE ROLE OF THE IMPLANTABLE LOOP RECORDER IN THE EVALUATION OF SYNCOPE

Nishant Gupta*, Shital Shah, Satish Tiyyagura and Mahesh Bikkina

Department of Cardiology, St. Joseph's Regional Medical Center, Paterson, NJ 07503.

ABSTRACT

Syncope is a transient and often un-witnessed event with numerous potential etiologies. There is often difficulty in choosing the appropriate diagnostic tests and procedures. Patients with risk factors for cardiac causes of syncope frequently undergo multiple testing modalities with poor yield and in one-third of patients, the cause remains unknown. Diagnosing cardiac causes of syncope is important as it portends a worse prognosis with a high recurrence rate and significant morbidity and mortality. The implantable loop recorder (ILR) is a small device that is implanted subcutaneously and can monitor cardiac rhythm. The ability of ILRs to continuously record cardiac electrical activity over extended periods of time makes them a powerful diagnostic tool. Several studies have now shown that the ILR can be very useful in the diagnostic evaluation of syncope and thus supports the recommendation that an ILR should be implanted early rather than late in the evaluation of infrequent unexplained syncope.

INTRODUCTION

Syncope is defined as a state of global cerebral hypoperfusion leading to transient loss of consciousness with associated loss of postural tone. It is characterized by rapid onset, short duration and spontaneous complete recovery without requiring electrical or chemical cardioversion and without neurological deficit. The addition of the term 'global cerebral hypo perfusion' in defining syncope helps distinguishing other causes of transient loss of

* Email: drnishant1702@gmail.com

consciousness like seizures , concussion and even stroke from the true causes of syncope. This gives a diagnostic and conceptual stability to the definition of syncope [1-4].

Syncope is a common medical problem that accounts for approximately 1% of medical admissions and 3% of emergency room visits. In elderly populations the annual incidence of syncope is 6%. Is has been postulated that this is due to the decrease in cerebral blood flow that accompanies the aging process. Patients with syncope also report markedly reduced quality of life. Based on the Framingham study, participants with an unknown cause of syncope had higher mortality rates as compared to those without syncope, however the participants with neurally mediated syncope did not experience increased rates of mortality [3]. The highest mortality was observed among those with cardiac cause of syncope. The fact that the population with unexplained causes of syncope had higher mortality as compared to those with neurally mediated or no syncope suggests that there might be fraction of patients with undiagnosed cardiac syncope in this pool. The findings in Framingham suggest that it is essential to identify and treat cardiac syncope so as to decrease mortality and improve outcomes. Arrhythmic syncope carries the worst prognosis and must be thoroughly evaluated. A correlation between symptoms and the cardiac rhythm is often difficult to determine during spontaneous syncope because of its sporadic, infrequent and unpredictable nature. In this chapter, we will first start with various causes of syncope and will then discuss the role of an extended monitoring strategy in patients with syncope[5-8].

APPROACH TO A PATIENT WITH SYNCOPE

Although a variety of diagnostic tests are available for the evaluation of syncope, a thorough history and physical examination is crucial to determine the cause and the best diagnostic approach. This will help to narrow the differential diagnosis of syncope which tends to be quite broad (Table 1). The history and physical can provide the diagnosis in up to 50% of cases. The initial evaluation should begin with an effort to ascertain that syncope actually occurred, rather than a disorder or insult that does not result in global cerebral hypoperfusion (Table 2).

The salient features in the history that suggest a cardiac arrhythmia include less than five seconds of warning and few or no prodromal symptoms. Demographic features suggesting an arrhythmia are male gender, less than three prior episodes of syncope and increased age.

On the other hand, a clinical history that favors a diagnosis of neurally mediated syncope includes palpitations, blurred vision, nausea, warmth, diaphoresis, or lightheadedness before the event and nausea, diaphoresis, or fatigue after the episode.

Special focus should be given to the following:

1) Whether the patient had a similar episode in the past.
2) Identifying the medications that may have a role.
3) Determine the history of cardiac disease or a family history of syncope or sudden cardiac death.
4) Investigate the type and duration of prodromal and recovery symptoms.

Table 1. Causes of Syncope

Reflex (neurally mediated) Syncope	Vasovagal/Neurocardiogenic Situational Carotid Sinus Syncope
Orthostatic Hypotension	Primary autonomic Failure Secondary autonomic failure Drug-induced Volume depletion
Cardiac Syncope: Arrhythmias: *Bradycardia*	Sinus node dysfunction Atrioventricular conduction defects Infrahisian conduction disease Implanted device malfunction
Tachycardia	Supraventricular Ventricular (idiopathic or secondary to channelopathies)
Structural Causes	Obstructive cardiac valvular disease Aortic dissection Atrial myxoma Pericardial disease/ tamponade Hypertrophic cardiomyopathy Myocardial ischemia or infarction Pulmonary embolus Pulmonary hypertension
Syncope of Unknown Origin	

Table 2. Conditions that Mimic Syncope

Disorders WITH loss of consciousness WITHOUT global hypoperfusion	Epilepsy Metabolic disorders Hypoglycemia Hypoxia Hyperventilation Intoxication Vertebrobasilar TIA
Disorders WITHOUT loss of consciousness	Cataplexy Drop attacks Falls Psychogenic Causes

Most importantly, it is very helpful to obtain careful accounts from witnesses who may have been present at the time of syncope.

The history should be followed by a careful physical examination. In examining a patient with syncope, it is essential to check blood pressure in both the arms as well as orthostatic blood pressure. In addition to a complete cardiac examination, assessing the patient's level of hydration, detecting the presence of significant neurological abnormalities suggestive of cerebrovascular accident or determining whether structural heart disease is present may provide relevant diagnostic clues.

Along with history, physical examination and orthostatic vital signs, the electrocardiogram (ECG) is part of an initial evaluation of syncope. Specific findings on the ECG that can suggest a possible cause of syncope include QT prolongation, a short PR interval and a delta wave (Wolff-Parkinson-White), right bundle branch block pattern with ST segment elevation (Brugada syndrome), high-grade AV block, T wave inversions in the right precordial leads (arrhythmogenic right ventricular cardiomyopathy), and the presence of Q waves. A normal ECG suggests a lower likelihood of a cardiac etiology of syncope in a young patient. Any abnormality of the baseline ECG is an independent predictor of cardiac syncope. In the patients with an initially normal evaluation, syncope is not associated with excess mortality. The main risk is associated with physical harm. A malignant episode of syncope is defined as an episode of syncope that occurs with little or no warning and results in a significant injury. Ambulatory ECG monitoring has an important role in these settings as the true arrhythmic episode may have been missed in the initial ECG which lasts only 10 seconds. The type and duration of ambulatory ECG monitoring is dictated by the frequency of symptoms. For example, the 24 to 48 hour holter monitor is appropriate for episodes that occur at almost daily. A typical event monitor allows ambulatory monitoring for 30 days and can be useful for when episodes are occurring at least once a month. These are small portable electrocardiographic recording devices that are usually worn at the belt with leads attached to the patient. They can be auto-triggered or can be activated by the patient. Continuous loop event monitors allows capture of both retrospective and prospective events. Patient triggering stores the ECG from 1 to 4 minutes before and 30 to 60 seconds after activation. These are recommended for the patients with infrequent but recurrent episodes of syncope. Recent studies have shown that event monitors have increased diagnostic yield as compared with holter monitoring. The limitations of ECGs, Holter monitors and event recorders are self-evident. If the syncopal event is infrequent, it is unlikely to occur during the time that the monitor device is worn [9-12].

THE ROLE OF IMPLANTABLE LOOP RECORDER IN SYNCOPE

For those patients with infrequent syncope and when arrhythmic causes are suspected, prolonged monitoring is the preferred method to make a diagnosis [13-16]. Since it is not practical to use a wearable monitor indefinitely, the best option is an implantable device that can remain in place long enough to document the next episode of syncope. The device is about a size of a memory stick, implanted subcutaneously in left anterior chest wall after a minor procedure, and requires no intra-cardiac electrodes (Figure 1).

Figure 1.

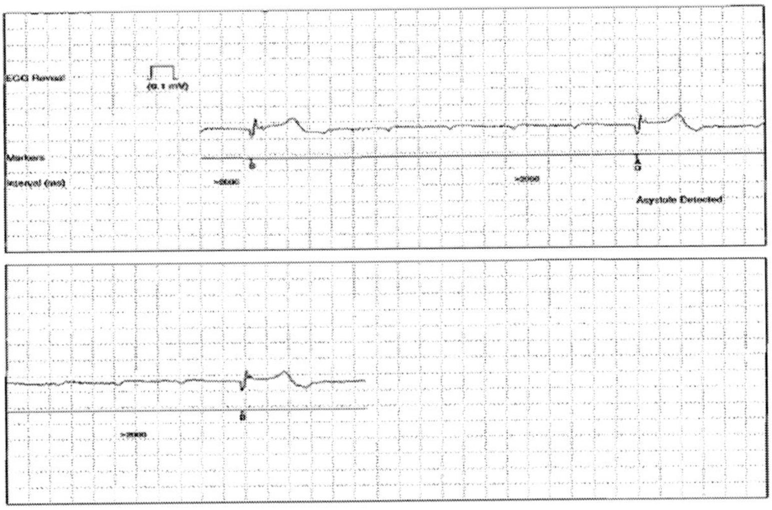

Figure 2.

A hand-held activator is provided to the patient, to trigger a recording when a symptom occurs. Additional features include auto activation for detected asystole, bradycardia, or tachyarrhythmia. Currently available ILRs have a battery life of up to 3 years. On the market in the United States, there are two implantable loop recorders: The REVEAL by Medtronic and the CONFIRM by St. Jude. The Medtronic ILR is the most prevalent and is a 61x19x8-mm recording device with 2 sensing bipoles 37 mm apart within its shell. The recorded ECG signal is stored in a circular buffer capable of retaining 21 minutes of uncompressed signal or 42 minutes of compressed signal in 1 or 3 divided parts. The memory buffer is frozen with a hand-held activator provided to the patient at the time of device implantation. The resultant programming can store a single episode of 21 minutes (20 minutes before and 1 minute after "freeze"), a 42-minute compressed episode (40 minutes before, 2 minutes after), three 7-minute episodes (6 minutes before, 1 minute after), or three 14-minute compressed episodes (12 minutes before, 2 minutes after). The episodes are then downloaded after interrogation with a standard programmer from the respective company.

The current approach to the patient with syncope involves a mixture of passive monitoring, which samples a relatively small time frame, and provocative testing aimed at actively reproducing the clinical situation. Both tilt-table and electrophysiological testing are negative in as many as 70% of cases and represent an artificial environment, making the interpretation of results particularly challenging [17-21]. The gold standard for establishing a diagnosis is the recording of physiological parameters during a spontaneous clinical episode. This makes the implantable loop recorder invaluable in well selected patients.

Multiple studies have been performed to determine the etiology of syncope. The landmark study was by Soteriades et al [22] and followed nearly 8000 patients for an average of 17 years as part of the Framingham Heart Study. During that time 822 patients developed syncope. Up to 37% of patients did not have a diagnosed cause of syncope. Neurocardiogenic syncope was diagnosed in 21%, cardiac causes in 10%, and neurologic causes in 9%. Mortality was highest in the patients with cardiac causes of syncope and in patients with neurocardiogenic syncope, there was no difference from the group without syncope. This study demonstrated the importance of identifying patients with cardiac causes of syncope.

One of the early studies using an implantable loop recorder was published in 1999 by Krahn et al [13]. The strategy of prolonged monitoring with an ILR was found to be very effective in establishing a cardiac diagnosis in patients with recurrent syncope. They selected 85 patients with recurrent undiagnosed syncope and used an ILR capable of cardiac monitoring for 18 months. During a mean follow up of 10.5+/_4 months, 68% of patients had symptom recurrence. An arrhythmia was detected in 42% of those patients. In a similar study by Nierop et al [23], arrhythmic cause for syncope was found in 23% of patients as compared to 28% by Lucas et al. This variation might be due to differing patient populations.

One large study which led to several sub-study publications was the International Study of Syncope of Uncertain Etiology (ISSUE), a multicenter international prospective study aimed at analyzing the diagnostic contribution of ILR in 4 predefined groups of patients with syncope of uncertain origin. The first group was the isolated syncope group patients without significant structural heart disease and normal tilt table testing The second group also had no significant structural heart disease but was tilt table test positive. The third group had bundle branch block and negative electrophysiological testing. The fourth group had suspected tachycardia because these were patients with previous myocardial infarction or cardiomyopathy with depressed ejection fraction or nonsustained ventricular tachycardia in whom an electrophysiological study did not induce sustained ventricular arrhythmias. The results of each patient population were reported separately below.

As described by Brignole and Menozzi, in patients with bundle branch block (BBB) and a negative EP study [24-25], most syncopal recurrences have a homogenous mechanism that is characterized by prolonged asystolic pauses mainly because of sudden-onset paroxysmal AV block. Indeed, one of my own patients fit this inclusion criteria and duplicated their findings 6 months after implantation (Figure 2). These findings are consistent with clinical feature of Stroke-Adams attacks. The prolonged asystole as the cause of syncope was in contrast to the above mentioned study by Krahn et al [18], where bradycardia was present in only a minority of patients. Similar results were also shown by Moya et al as part of the ISSUE Trial (International Study on Syncope of Uncertain Etiology), where they found that recurrent syncopal episodes in patients with no structural heart disease and normal EKGs were due to sinus arrest. In this trial, ILR documentation of syncope recurrence was obtained in 32 of 111 patients (29%). Similarly, in another subset of the ISSUE study by Menozzi et al, out of 35 patients with overt heart disease, 17% had ILR-documented syncope during follow up. As mentioned above, a third ISSUE report by Brignole et al showed progressive AV Block as the cause of recurrent syncope in 36% of patients with BBB and negative EP study [24-26].

In a recent prospective , multicenter , observational PICTURE (Place of Reveal In the Care pathway and Treatment of patients with Unexplained Recurrent Syncope) registry [27], out of 218 patients with recurrent syncopal events , an ILR- guided diagnosis was obtained in 170 cases (78%), and of the 170, 128 (75%) were cardiac. In this registry, the recurrence of syncope was 19%, 26%, and 36% after 3, 6, and 12 months, respectively. It was noted that a longer duration of follow up to >1yr increases the chance of recurrence of syncope. The recurrence rate of syncope post implantation dictates the success rate of ILR monitoring. Farwell and Sulke [17,28] also noted enhanced diagnosis rates with an increased use of automatic recording of events. Half the population in PICTURE received the Reveal (Medtronic ILR) device models where the automatic recording mode is always active. In the other half of the patients, when automatic detection was not the default and could be switched off at the discretion of the investigator, we cannot exclude that the yield would have been improved if the automatic detection had always been enabled. Interestingly, in the PICTURE registry, the proportion of patients with an ILR-guided diagnosis increased over time and was still growing at the end of follow-up, implying that more patients would get a diagnosis before the end of battery life. These findings favor the current guidelines which recommended placing an ILR sooner rather than later for the evaluation of unexplained recurrent syncope as it has significant impact on the overall prognosis [29].

The large body of evidence in favor of implantable loop recorders has led to the European society of cardiology guidelines on management of Syncope to include the use of implantable loop recorders early in the management of syncope as paraphrased below [29-30].

Indications

- In the early phase of evaluation in patient with recurrent syncope of uncertain origin, the absence of high risk criteria and a high likelihood of recurrence within battery longevity of device.
- High risk patients in whom a complete evaluation did not demonstrate a cause of syncope or lead to a specific treatment.
- ILR should be considered to assess the contribution of bradycardia before embarking on cardiac pacing in patients with suspected or certain reflex syncope presenting with frequent or traumatic syncopal episodes.

Diagnostic Criteria

- EKG monitoring is diagnostic when a correlation between syncope and an arrhythmia (brady- or tachyarrhythmia) is detected.
- In the absence of such correlation, EKG monitoring is diagnostic when periods of Mobitz II or III degree AV block or a ventricular pause >3s (with exceptions: young trained persons, during sleep, medicated patients, or rate controlled atrial fibrillation) or rapid prolonged paroxysmal SVT or VT are detected. The absence of arrhythmia during syncope excludes arrhythmic syncope.
- The EKG documentation of pre-syncope without any relevant arrhythmia is not an accurate surrogate for syncope.
- Asymptomatic arrhythmias (other than those listed above) are not an accurate surrogate for syncope.
- Sinus bradycardia in the absence of syncope is not an accurate surrogate for syncope.

Conclusion

The implantable loop recorder clearly acts a powerful tool for establishing the relationship between syncope and arrhythmias in patients with infrequent unexplained syncope. The body of literature to support their use in syncope has been accumulating for over 10 years and is sufficient to support their use early in the management of syncope and may reduce the use of less cost-effective testing.

REFERENCES

[1] Manolis AS, Linzer M, Salem D, Estes NA III. Syncope: current diagnostic evaluation and management. *Ann. Intern. Med.* 1990; 112: 850–863.

[2] Manolis AS. The clinical spectrum and diagnosis of syncope. *Herz.* 1993; 18: 143–154.

[3] Kapoor WN, Karpf M, Wieand S, Peterson JR, Levey GS. A prospective evaluation and follow-up of patients with syncope. *N. Engl. J. Med.* 1983; 309: 197–204.

[4] Shen WK, Decker WW, Smars PA, Goyal DG, Walker AE, Hodge DO, Trusty JM, Brekke KM, Jahangir A, Brady PA, Munger TM, Gersh BJ, Hammill SC, Frye RL. Syncope Evaluation in the Emergency Department Study (SEEDS): *Circulation.* 2004 Dec 14;110(24):3636-45. Epub 2004 Nov 9.

[5] Brignole M, Ungar A, Bartoletti A, Ponassi I, Lagi A, Mussi C, Ribani MA, Tava G, Disertori M, Quartieri F, Alboni P, Raviele A, Ammirati F, Scivales A, De Santo T; Evaluation of Guidelines in Syncope Study 2 (EGSYS-2) GROUP. Standardized-care pathway vs. usual management of syncope patients presenting as emergencies at general hospitals. *Europace.* 2006 Aug;8(8):644-50.

[6] Kapoor WN. Evaluation and outcome of patients with syncope. *Med. (Balt).* 1990; 69: 160–175.

[7] Day SC, Cook EF, Funkenstein H, Goldman L. Evaluation and outcome of emergency room patients with transient loss of consciousness. *Am. J. Med.* 1982; 73: 15–23.

[8] M.J. Reed, D.E. Newby, A.J. Coull, K.G. Jacques, R.J. Prescott and A.J. Gray, The Risk stratification Of Syncope in the Emergency department (ROSE) pilot study: a comparison of existing syncope guidelines, *Emergency Med. J.* 24 (2007), pp. 270–275.

[9] Benditt DG, Nguyen JT. Syncope: therapeutic approaches. J Am Coll Cardiol. 2009 May 12;53(19):1741-51.

[10] Boersma L, Mont L, Sionis A, Garcia E, Brugada J. Value of implantable loop recorder for the management of patients with unexplained syncope. *Europace* 2004;6:70–76.

[11] Pierre B, Fauchier L, Breard L, Marie O, Poret F, Babuty D. Implantable loop recorder for recurrent syncope: influence of cardiac conduction abnormalities showing up on resting electrocardiogram and of underlying cardiac disease on follow-up developments. *Europace* 2008;10:477–481.

[12] Krahn AD, Klein GJ, Yee R, Skanes AC. Randomized assessment of syncope trial: conventional diagnostic testing versus a prolonged monitoring strategy. *Circulation.* 2001 Jul 3;104(1):46-51.

[13] Krahn AD, Klein GJ, Yee R, Takle-Newhouse T, Norris C. Use of an extended monitoring strategy in patients with problematic syncope. Reveal Investigators. *Circulation* 1999;99:406-10.

[14] Lombardi F, Calosso E, Mascioli G, Marangoni E, Donato A, Rossi S, et al. Utility of implantable loop recorder (Reveal Plus) in the diagnosis of unexplained syncope. *Europace* 2005;7:19-24.

[15] Krahn AD, Klein GJ, Yee R, Skanes AC. Detection of asymptomatic arrhythmias in unexplained syncope. *Am. Heart J.* 2004;148:326-32.

[16] Solano A, Menozzi C, Maggi R, Donateo P, Bottoni N, Lolli G, et al. Incidence, diagnostic yield and safety of the implantable loop-recorder to detect the mechanism of syncope in patients with and without structural heart disease. *Eur. Heart J.* 2004;25:1116-9.

[17] Farwell DJ, Sulke AN. Does the use of a syncope diagnostic protocol improve the investigation and management of syncope? *Heart* 2004;90:52-8.

[18] Krahn A, Klein G, Norris C, Yee R. The etiology of syncope in patients with negative tilt table and electrophysiologic testing. *Circulation* 1995;92:1819–1824.

[19] Ermis C, Zhu AX, Pham S, Li JM, Guerrero M, Vrudney A, Hiltner L, Lu F,Sakaguchi S, Lurie KG, Benditt DG. Comparison of automatic and patient activated arrhythmia recordings by implantable loop recorders in the evaluation of syncope. *Am. J. Cardiol.* 2003;92:815–819.

[20] Moya A, Brignole M, Menozzi C, Garcia-Civera R, Tognarini S, Mont L, Botto G,Giada F, Cornacchia D. Mechanism of syncope in patients with isolated syncope and in patients with tilt-positive syncope. *Circulation* 2001;104:1261–1267.

[21] Moya A, Brignole M, Sutton R, Menozzi C, Garcia-Civera R, Wieling W, Andresen D, Benditt D, Garcia-Sacristan JF, Beiras X, Grovale N, De Santo T, Vardas P. Reproducibility of electrocardiographic findings in patients with neurally-mediated syncope. *Am. J. Cardiol.* 2008; 102:1518–1523.

[22] Soteriades ES, Evans JC, Larson MG, Chen MH, Chen L, Benjamin EJ, Levy D. Incidence and prognosis of syncope. *N. Engl. J. Med.* 2002; 347: 878–885.

[23] Nierop P, van Mechelen R, Elsacker A, Luijten RH, Elhendy A. Heart rhythm during syncope and presyncope: results of implantable loop recorders. *Pacing Clin. Electrophysiol.* 2000;23:1532–1538.

[24] Brignole M, Menozzi C, Moya A, Garcia-Civera R, Mont L, Alvarez M, Errazquin F, Beiras J, Bottoni N, Donateo P; International Study on Syncope of Uncertain Etiology (ISSUE) Investigators. Mechanism of syncope in patients with bundle branch block and negative electrophysiological test. *Circulation* 2001;104:2045–2050.

[25] Menozzi C, Brignole M, Garcia-Civera R, Moya A, Botto G, Tercedor L, Migliorini R, Navarro X; International Study on Syncope of Uncertain Etiology (ISSUE) Investigators. Mechanism of syncope in patients with heart disease and negative electrophysiologic test. *Circulation* 2002;105:2741–2745.

[26] Brignole M, Sutton R, Menozzi C, Garcia-Civera R, Moya A, Wieling W, Andresen D, Benditt DG, Vardas P; International Study on Syncope of Uncertain Etiology 2 (ISSUE 2) Group. Early application of an implantable loop recorder allows effective specific therapy in patients with recurrent suspected neurally mediated syncope. *Eur. Heart J.* 2006 May;27(9):1085-92. Epub 2006 Mar 28.

[27] Edvardsson N, Frykman V, van Mechelen R, Mitro P, Mohii-Oskarsson A, Pasquié JL, Ramanna H, Schwertfeger F, Ventura R, Voulgaraki D, Garutti C, Stolt P, Linker NJ; PICTURE Study Investigators. Use of an implantable loop recorder to increase the diagnostic yield in unexplained syncope: results from the PICTURE registry. *Europace.* 2011 Feb;13(2):262-9. Epub 2010 Nov 19.

[28] D.J. Farwell, N. Freemantle and A.N. Sulke, Use of implantable loop recorders in the diagnosis and management of syncope, *Eur. Heart J.* 25 (2004), pp. 1257–1263.

[29] Task Force for the Diagnosis and Management of Syncope; European Society of Cardiology (ESC); European Heart Rhythm Association (EHRA); Heart Failure Association (HFA); Heart Rhythm Society (HRS), Moya A, Sutton R, Ammirati F, Blanc JJ, Brignole M, Dahm JB, Deharo JC, Gajek J, Gjesdal K, Krahn A, Massin M, Pepi M, Pezawas T, Ruiz Granell R, Sarasin F, Ungar A, van Dijk JG, Walma EP, Wieling W. Guidelines for the diagnosis and management of syncope (version 2009). *Eur. Heart J.* 2009 Nov;30(21):2631-71. Epub 2009 Aug 27.

[30] Task Force members, Brignole M, Vardas P, Hoffman E, Huikuri H, Moya A, Ricci R, Sulke N, Wieling W; EHRA Scientific Documents Committee, Auricchio A, Lip GY, Almendral J, Kirchhof P, Aliot E, Gasparini M, Braunschweig F; Document Reviewers, Lip GY, Almendral J, Kirchhof P, Botto GL; EHRA Scientific Documents Committee. Indications for the use of diagnostic implantable and external ECG loop recorders. *Europace.* 2009 May;11(5):671-87.

INDEX

D

E

F

G

H

I

J

L

M

Q

R

T

U

V

W

Y